Leaving Fossil Fuels Underground

Leaving Fossil Fuels Underground

Actors, Arguments and Approaches in the Global South and Global North

Edited by
Joyeeta Gupta,
Barbara Hogenboom,
Arthur Rempel and
Malin Olofsson

Routledge
Taylor & Francis Group
LONDON AND NEW YORK

This project was partly financed by the Netherlands Scientific Foundation NWO (project W07.303.104 [2017–2023]); the European Research Council (ERC) Advanced Grant under the European Union's Horizon 2020 research and innovation programme (grant agreement no. 101020082 [2021–2026]) on Climate Change and Fossil Fuel) and partly by the Governance and Inclusive Development programme group, the Centre for Latin American Studies and Documentation (CEDLA) and the Faculty of Humanities of the University of Amsterdam.

First published in 2024 by Amsterdam University Press Ltd.

Published 2025 by Routledge
4 Park Square, Milton Park, Abingdon, Oxon OX14 4RN
605 Third Avenue, New York, NY 10158

Routledge is an imprint of the Taylor & Francis Group, an informa business

ISBN: 9789048560370 (hbk)
ISBN: 9781041182085 (pbk)
ISBN: 9781003698838 (ebk)
NUR 600

Cover design: Coördesign, Leiden

DOI: 10.5117/9789048560370

For Product Safety Concerns and Information please contact our EU representative:
GPSR@taylorandfrancis.com
Taylor & Francis Verlag GmbH, Kaufingerstraße 24, 80331 München, Germany

Table of Contents

List of Tables

List of Figures

List of Boxes

List of Abbreviations

AGM	annual general meeting
ANC	African National Congress
BDB	bilateral development bank
CBAM	Carbon Border Adjustment Mechanism
CBO	community-based organisation
CCL	climate change litigation
CJC	Climate Justice Coalition
CJCM	Climate Justice Charter Movement
COP	Conference of the Parties
COSATU	Congress of South African Trade Unions
DFFE	Department of Forestry, Fisheries and the Environment
DMRE	Department of Mineral Resources and Energy
ECA	export credit agency
EIA	environmental impact assessment
EIUG	Energy Intensive Users Group of Southern Africa
EJAtlas	Global Atlas of Environmental Justice
EU	European Union
GDP	gross domestic product
GHG	greenhouse gas
IACHR	Inter-American Court of Human Rights
INGO	international non-governmental organisation
IPCC	Intergovernmental Panel on Climate Change
IPP	independent power producer
IRP	Integrated Resource Plan
JET IP	Just Energy Transition Investment Plan
JETP	Just Energy Transition Partnership
LFFU	leaving fossil fuels underground
LNG	liquefied natural gas
LOS	legal opportunity structure
MA	mitigation alliance
MACUA	Mining Affected Communities United in Action
MDB	multilateral development bank
MEC	minerals–energy complex
MMSEZ	Musina-Makhado Special Economic Zone
NCP	nature's contributions to people
NDC	nationally determined contribution
NGO	non-governmental organisation

NIMBY	"not in my back yard"
NUMSA	National Union of Metalworkers of South Africa
NZAOA	Net-Zero Asset Owner Alliance
OECD	Organisation for Economic Co-operation and Development
OPEC	Organization of the Petroleum Exporting Countries
PA	Paris Agreement
PACJA	Pan African Climate Justice Alliance
PFI	public finance institution
SA	South Africa
SAFTU	South African Federation of Trade Unions
SCC	social cost of carbon
SDG	Sustainable Development Goal
SEZ	special economic zone
SRI	socially responsible investment
UKEF	UK Export Finance
UN	United Nations
UNCED	United Nations Conference on Environment and Development
UNDP	United Nations Development Programme
UNEP	United Nations Environment Programme
UNESCO	United Nations Educational, Scientific and Cultural Organization
UNFCCC	United Nations Framework Convention on Climate Change
UNGA	United Nations General Assembly
WAMUA	Women Affected by Mining United in Action
WWF	World Wildlife Fund
YNP	Yasuní National Park

Foreword and Acknowledgements

Thirty years of research into climate change has revealed that a topic that is scarcely discussed is the constituency of vested interests in the fossil fuel sector that has played a strong role in delaying action on climate change. This is because the vested interests range from large investors to small fossil fuel users, from states in fossil fuel-extracting countries to states in importing countries, and from the Global North to the Global South. Hence, this book focuses on identifying a kaleidoscope of important ideas and initiatives around leaving fossil fuels underground (LFFU) in the special context of the developing world.

This book arises from a collaboration between scholars at the University of Amsterdam and those in South Africa and Ecuador for the project "Leave Fossil Fuels Underground for Sustainable and Inclusive Development: Co-creating Alternative Pathways in Africa and Latin America." The collaboration builds on the ongoing research work of the project leaders (Barbara Hogenboom and Joyeeta Gupta), draws inspiration from and builds on collaborative efforts with Patrick Bond of South Africa and Carlos Larrea of Ecuador and centres around the work of two PhD scholars—one from Ecuador (Carolina Valladares) and the other from Italy/Brazil who studied South Africa (Arthur Rempel). It also builds on contact with Oilwatch Latin America and on the work of 20 MSc scholars[1] who have worked on a range of related issues as part of this project. Three interns have also supported the research and outreach work—Benjamin Gramsch, Lisa Pier and Aya Wietzorrek. In addition, Nina Brander, Hilmer Bosch and Viktoria Ruseva have contributed to editing, and Kimberley Williams has made valuable contributions to updating some of the chapters and writing the text for the boxes. We were also supported by an online network of scholars and activists called the Euro-African Community for Leaving Fossil Fuels Underground (EAC-LFFU). This project was partly financed by the Dutch Research Council (Nederlandse Organisatie voor Wetenschappelijk Onderzoek, NWO) (project W07.303.104) and partly by the Governance and Inclusive Development programme group, the Centre for Latin American Studies and

1 Anne Politsch (2019), Katarina Sawicka-Wrzask (2019), Joris den Breejen (2019), Mariv Pront (2019), Arthur Rempel (2019), Niels Moek (2019), Adrien Tofighi (2019), Emma McCarthy (2020), Tjeerd Harkema (2020), Aaron Baard (2020), Isabelle Hilson (2020), Nina Brander (2020), Jade Saint-Rose (2020), Clara McDonnel (2021), Jennifer van Beek (2021), Patrick Higgins (2021), Marit Kerckhoffs (2021), Damian van Slooten (2021), Alfie Begley (2021) and Laura Johnson (2021).

Documentation (CEDLA) and the Faculty of Humanities of the University of Amsterdam. This work is now being followed-up further by Joyeeta Gupta in the European Research Council (ERC) Advanced Grant under the European Union's Horizon 2020 research and innovation programme (grant agreement no. 101020082) on Climate Change and Fossil Fuel. Part of this book has been written under the latter project and this forms a base line of thoughts regarding how this problem can be systematically addressed.

Two of the critical years for fieldwork for this project were affected by the COVID-19 pandemic and this has led to a number of challenges with respect to actually setting up a dialogue with people in the field to co-produce knowledge. However, our monthly online dialogues with colleagues in South Africa have provided an interesting process and venue for discussion on how to address the challenge of phasing out fossil fuels in Africa, ideas and issues that are also relevant for Latin America. In addition, our project's meeting in Ecuador in August 2018 and the meeting in South Africa in July 2022 allowed for valuable encounters with a wide range of local actors and experts and field trips to sights of LFFU struggles.

Providing a systematic and fully integrated appraisal of all the key issues at play when leaving fossil fuels underground is beyond the scope of this volume. Rather, through analyses of a range of interesting LFFU ideas and experiments as well as the bottlenecks for upscaling, which are loosely linked to our analytical approach anchored in environmental justice, inclusive development and transition theory, we hope this book inspires much needed further research, debate, action and policy to leave fossil fuels underground. At the time of writing this foreword, a new global crisis induced by the Russian attack on Ukraine is not only hurting globally the political, economic and financial domains but also access to energy and (thereby) food in both the South and North. This situation poses new bottlenecks that need to be dealt with in order to achieve the energy transition, socio-environmental justice and inclusive development.

Joyeeta Gupta, Barbara Hogenboom, Arthur Rempel and Malin Olofsson
Amsterdam, June 2024

1. Leaving Fossil Fuels Underground

Joyeeta Gupta, Barbara Hogenboom, Arthur Rempel, Carolina Valladares and Hebe Verrest

Abstract
How do different actors engage with fossil fuel phase-out? Which arguments and approaches are successful in dealing with the multiple trade-offs involved in the fossil fuel discussion, and how can these be scaled up? And, most fundamentally, how can societies move away from a global system based on fossil fuels and its deeply vested economic, financial and political interests and achieve energy transition, inclusive development and socio-environmental justice? This introductory chapter points at the academic, societal and policy importance of understanding and engaging with leaving fossil fuels underground (LFFU) initiatives. It presents the knowledge gap regarding the ways in which different actors, ranging from investors to social movements and from state to non-state actors at all levels of the geographical scale, use and develop different arguments and approaches in phasing out fossil fuels at multiple levels of governance. Finally, it explains how the other chapters in the book examine the potential of LFFU within a global framework of North–South studies, with special reference to Africa and Latin America and case studies on South Africa and Ecuador.

Keywords: fossil fuel phase-out, inclusive development, socio-environmental justice, governance, energy transition

1.1. The purpose of this book

In order to address the global climate emergency (European Parliament, 2019), it is increasingly important to bring net greenhouse gas (GHG) emissions to zero by 2050. The Paris Agreement on Climate Change (Paris

---

Gupta, J., Hogenboom, B., Rempel, A. & Olofsson, M. (eds), *Leaving Fossil Fuels Underground: Actors, Arguments and Approaches in the Global South and Global North.*
Taylor & Francis Group, 2024
doi: 10.5117/9789048560370_CH01

Agreement, 2015) (see 1.5) has been ratified or acceded to by almost all countries (190 states, including China and the United States, and the EU). It calls for: (a) the limitation of global temperature rise to a maximum of 2° above pre-industrial levels, and if possible to be kept as low as 1.5 °C; (b) the strengthened capacity of countries to manage the impacts of climate change; and (c) the alignment of financial flows with these goals to support the mitigation of climate change and the actions of developing countries (Paris Agreement, 2015). This science-based target of the Paris Agreement supports grassroots action against fossil fuel extraction and use because of the local-level challenges associated with fossil fuels. At the same time, it requires that actors, at all levels of governance, take action to jointly work toward leaving fossil fuels underground (LFFU)—up to 60% of proven oil and gas reserves and 90% of coal reserves to comply with the 1.5 °C goal (Welsby et al., 2021). At the same time, taking measures in relation to the Paris Agreement raises all kinds of important justice issues—justice issues in relation to extraction, climate impacts, stranded resources and assets, affordable energy, contract and investment law and so on. We explore these justice issues throughout the book and return to the broader question of what climate change implies for development in the final chapter.

Against this background, this book examines the role of key actors, arguments and approaches in promoting the rapid phase-out of fossil fuels. Since these LFFU actors, arguments and approaches can be found at and across all geographical levels, this book has a multi-level perspective, with a special focus on South Africa in Africa and Ecuador in Latin America. A key factor is those who resist the use of fossil fuels for varying reasons. While most of the early resistance against fossil fuel extraction happened at the local level by affected groups, increasingly, these actors have collaborated with other groups and institutions as well as scholars at the national and transnational level. Further, with climate change impacts and knowledge expanding, national and global action have expanded too. Other actors include those who invest in fossil fuels and have major vested interests in the sector; those who use, consume and depend on fossil fuels for their livelihoods and businesses and these investment, production and consumption actors are also major players.

Despite 30 years of international climate negotiations and the Paris Agreement, numerous national policy reforms and the start of a global energy transition, the current pace of phasing out fossil fuels is far too slow to limit global temperature rise to 1.5–2.0 °C (Aengenheyster et al., 2018). In fact, industrialised countries have continued to exploit new fossil fuel reserves and many countries in the Global South follow suit by engaging in large new fossil fuel projects and related debts, with the support of institutions from the Global North, instead of developing new economic and energy strategies. In response, non-state actors and social movements are mobilising to address various aspects of the fossil fuel problem. Moreover, social, academic and political debates increasingly question how societies can move away from a century-old global system based on almost unlimited use of fossil fuels, and the deeply vested economic, financial and political interests and patterns. This book contributes to this debate, by exploring proposals and initiatives for new and coherent international policies and financial mechanisms for LFFU.

1.2. Scientific relevance: Why leave fossil fuels underground?

A key question is: Why are we focusing on LFFU as opposed to a more general narrative on how does one promote the energy transition? This is because we have identified a set of important knowledge gaps around four interrelated areas: First, the difficulties and opportunities in phasing out fossil fuel extraction; second, the need for coherence in related international financial flows; third, while there is considerable discussion on the North–South dynamics of emission sharing, there is very little that examines this from the perspective of who is allowed to use the remaining fossil fuel and the right to development; and fourth, the fact that in a globalised world, decisions to phase out fossil fuels involve actors from all continents and at all levels of governance. We argue that there is little comprehensive analysis of how different actors, ranging from investors to social movements and from state to non-state actors at all levels of the geographical scale, use and develop different arguments and approaches in phasing out fossil fuel at multiple levels of governance. We briefly explain these gaps in knowledge below.

First, we aim to address the "background" difficulties and opportunities in phasing out fossil fuels, which have not received sufficient scholarly attention. The Paris Agreement implies a global energy transition to renewables and non-fossil energy. Considerable work is taking place worldwide to assess the potential of such a transition (see Chapter 2). In order to promote such a transition, fossil fuels and the fossil fuel economy have to be phased out to make space for a new and revamped renewable and non-fossil-fuel energy economy and this can possibly only happen within a very different development context (see 1.4). Such an energy transition requires, on the one hand, losing revenues from stranded assets and resources (see Chapter 2) of €14.27 ($16) trillion to €263 ($295) trillion (Linquiti & Cogswell, 2016) (depending on the methods used to compute the losses) globally while requiring an investment of $115 trillion in low-carbon and renewable energy technologies by 2050. This also implies phasing out all existing fossil fuel infrastructure (e.g. pipes for transporting gas by electricity lines) and technology and replacing that with new approaches—for example, removing all fossil fuel-based cooking by electric cooking, or all fossil fuel-based transport by electric transport.

However, while there is considerable research on the energy transition and the shift from coal to oil to gas, there is inadequate research on why it is so politically and economically difficult to globally leave fossil fuels underground at the fast pace that the climate emergency requires: the abundance of global fossil fuel reserves; the substantial financial resources invested in the sector; the nature of infrastructural and technological dependency on the fossil fuel sector; the political, financial and technological lock-in; and how these all interact. Phasing out fossil fuels could lead to major risks for the financial sector as the European Investment Bank has shown and even the collapse of the financial sector (Clark, 2020, citing Mike Carney, former governor of the Bank of England). This is because the fossil fuel enterprise is valued at a higher than realistic price, creating a carbon bubble (Carbon Tracker Initiative, n.d.). This carbon bubble implies that the economic value of fossil fuel assets is considerably more than it would be if the implications of the Paris Agreement had been factored in. Those who see these assets on their balance sheets are unwilling to see them evaporate if they were to become stranded assets or stranded resources (Bos & Gupta, 2019; see Chapter 2 for explanation of stranded assets).[1] Many individuals, companies and governments are still hedging their bets and hope that they can maximise their individual profits before a fossil fuel phase-out becomes legally mandated. Ongoing investments in fossil fuels today may lead to emissions that result in average global temperatures to rise to 1.5 °C above preindustrial levels by 2040 (Millar et al., 2018), while more recent projections see us crossing 1.5 °C by 2030.

It is important to point out that the Paris Agreement does not mention the word "fossil fuel" and thus does not directly address these actors and interests. In a way, the global economics and politics of a fossil fuel phase-out

1 For an individual, a stranded asset (e.g. a fossil fuel-using car) implies the costs of replacement or the lack of employment for those employed in the sector. For a company, a stranded asset implies loss of shareholder value (Carbon Tracker Initiative, n.d.). For an economy, a stranded asset implies the potential risk of economic collapse (Mercure et al., 2018). These costs lead to resistance from fossil fuel users, producers and investors from household to global level (Bos & Gupta, 2019).

is the *background* against which decisions are made in international treaty negotiations and national governments. This crucial background, however, has been inadequately studied in the scholarly literature, which has mostly focused on the *foreground* of the climate negotiations and the implementation challenges of the individual articles (Gupta, 2014). At the same time there are many opportunities to phase out fossil fuels, which have created major environmental, economic, social and political externalities, not only at the global and national level—massive pollution, Dutch disease, boom–bust development and corruption, to name a few—but also at the local level, from extraction processes in rural areas to use in urban areas. This has mobilised large groups of actors against the use of fossil fuel, in particular at the local level; but equally at the local level there are actors that resist the phase-out of fossil fuel since they depend on it for their livelihoods.

Second, a related gap in knowledge concerns the incoherence of fossil fuel-related international financial flows with the Paris Agreement and climate change mitigation. The fossil fuel sector is a global sector. There are huge global financial flows with respect to it. There is prima facie evidence that these flows are not consistent with the mitigation of climate change. Massive financial resources continue to be used to discover new sources of fossil fuels and to invest in them (Carbon Tracker Initiative, 2019; Gupta et al., 2020). The G20 governments have been noted to continue supporting the fossil fuel sector, and even propping up the sector during the COVID-19 crises (Geddes et al. 2020; Rempel & Gupta, 2021). This is happening probably because those who are divesting their investments are selling these to the Global South (see Chapter 2). It may also happen because the industry is sensing a crisis in the Global North and is finding ways to shift the responsibility to the Global South, but these hypotheses need testing. This is why the Paris Agreement of 2015 called for coherence in financial flows, as the lack of consistency further entrenches fossil fuel interests and creates new vested interests.

Third, and in relation to the above two gaps in knowledge, is the North–South dimension of the global fossil fuel dilemma. The literature has identified the tensions in globally sharing the carbon budget (the maximum amount of carbon that can be emitted if we want to achieve a certain temperature level) in a fair way, in relation to the size of the budget, the link to development levels, the temporal aspect of whether one can postpone decision-making in the hope of new technologies resolving these tensions, the question of how to share the budget between countries and peoples, the impact of trade on the budget (should one focus on production-related emissions or consumption-related emissions) and the issue of whether such a transition is actually possible given the limits and challenges to non-fossil-fuel-related emissions (Gupta, 2014). However, the literature has not covered the relationship between (a) sharing emissions and sharing the right to use natural resources within one's own country; (b) the tensions between the right to development and the right to (promote) sustainable development (see Chapter 2); and (c) how the incoherence in financial flows affects North–South relations. While there is a huge literature focusing on the issue of dividing or sharing emission rights between countries, these are not embedded in an understanding of the global fossil fuel dilemma and its different North–South dimensions (see 1.3). Achieving the 1.5 °C goal will alter global, national and local financial and energy sectors and thereby hinder the development prospects of the developing world, undermining their ability to ensure sustainable livelihoods unless they can redefine their development paradigm and internalise ecological damage (Gupta, 1997; Hicks et al., 2008; Okereke & Coventry, 2016).

Climate change, therefore, is not just a serious environmental issue, but it is, in essence, a development issue (Gupta & Van der Grijp, 2010, p. xiv). Climate change shows us that we need to live within planetary boundaries and thus fundamentally questions our understanding and definition of development. The North–South tensions have been exacerbated by the fact that although the original idea of the UN Framework Convention on Climate Change (UNFCCC, 1992) was that the rich countries would reduce their emissions to make space for the developing countries, over the years from 1990 to 2020, there has only been a period of five years (2008–2012) during which some of the rich countries (excluding the US and Canada) had legally binding targets. The Doha Amendment to the Kyoto Protocol (2012) only entered into force on the last day of the 2020, thus undermining its effectiveness. The Paris Agreement has in turn entered the post-equity stage where every country individually does what it can and has come under much critique from equity scholars (see 1.5.1). This is all the more poignant given that a substantial part of the remaining fossil fuel reserves is in the Global South. These countries will de facto be left with stranded resources and assets as the bulk of the carbon budget has been used up by the major industrialised countries, with China and India trying to catch up (Bos & Gupta, 2018). It is possible that the Global South will not only have to face the heaviest impacts of climate change to which their contribution has been relatively small; but I can also not use the same development path as used by the Global North, nor can it aspire to the production and consumption patterns of the Global North, which are completely unsustainable (Gupta, 2014). Equitably phasing out fossil fuels thus requires reimagining alternative pathways to development beyond the orthodox growth paradigms.

Fourth, and following from the above, while there is vast scholarship on the role of states, leaders, businesses and non-governmental organisations (NGOs) in the area of climate change, there is very little dedicated scholarship on the role of those actors that are critical for phasing out fossil fuels and that function in the "background." For example, there is growing evidence that about half of the large (parent) companies along the global coal value chain plan to start new coal mines, plants and infrastructure, and will be expanding their operations as long as shareholders and institutional investors continue to offer the financial resources to do so (Urgewald, 2020). More generally, fossil fuel interests are promoted by fossil fuel companies and those who invest in them, including the aid and export credit agencies (Gupta et al., 2020), pension funds (Rempel & Gupta, 2020), development and commercial banks and philanthropic foundations. Some multilateral institutions such as the European Investment Bank have pledged to phase out fossil fuel investments by 2021 (EIB, 2019) and the World Bank has promised to stop investments in oil and gas exploration (World Bank, 2019). However, most other banks across the world have not.

It is not just the big investors that are not covered adequately by the literature; it is also the social movements. These are growing worldwide and range from those opposing fossil fuel extraction because of the local impacts to those opposing climate change (Fridays for Future, Extinction Rebellion). Coal, oil and gas extraction is facing opposition because of the multifaceted problems it propagates in addition to climate change: water and land grabbing, local water, soil and air pollution, deforestation, health damage, unfarmable land and famine, and repression via neocolonialism and environmental racism (Urgewald, 2020). Focused on the oil sector, Watts (1999) has coined this multifaceted destruction and damage as "petro-violence." The system of the extraction, transport, refining and consumption of oil and other fossil fuels that directly and indirectly causes widespread damage has relevant similarities to the fast and slow violence and the widespread harm of the system of colonialism (McDermott Hughes, 2017).

While there is substantial literature on individual case studies of individual social movements resisting fossil fuel industries, there is little that presents an overview of the different actors, their arguments and the approaches to promote or hinder the process of leaving fossil fuels underground. This is all the more important as halting climate change and leaving fossil fuels underground is a "super wicked" problem: the gains and losses faced by different actors vary, the stakes are very high and those who bear the costs define the problem differently from those who benefit.

Thus, the above four interrelated gaps in knowledge—the huge financial aspects of the fossil fuel sector, the incoherence in funding flows, the evolving

North–South dimensions and the role of actors, arguments and approaches in this context are the major focus of this book. All these approaches imply huge trade-offs for the different actors engaged in the debate and this has implications for how they engage. A key challenge is whether we can move beyond these trade-offs to identify successful LFFU strategies that are socially, ecologically and politically inclusive (see 2.2).

1.3. Research focus and limits

Given the above gaps in scholarship, this book focuses on the question: How do different actors use different arguments and approaches to engage with fossil fuel phase-out? Which arguments and approaches are successful in dealing with the multiple trade-offs involved in the fossil fuel discussion for which actors and why? And how can these be scaled up? Embedded within a global framework of North–South studies, these questions are answered with special reference to Africa and Latin America (see Chapters 4–8). We examine LFFU potential and problems from a bottom-up approach and top-down approach. By partly focusing on the cases of South Africa and Ecuador, we can include insights from dynamics at the local and national levels into our analysis. We look at the social movements that are protesting the fossil fuel sector locally as well as nationally, regionally and globally. Then we zoom into the role of states who are often caught between their responsibilities to respond to contestation of existing policies and practices, and their fear that alternative approaches may impact on their economic growth and the vested interests. Next, we look at the actors who invest in the fossil fuel sector. An assessment of the actors, arguments and approaches within the fossil fuel sector itself (private fossil fuel companies and state-owned fossil fuel companies) is beyond the scope of this book. Being the world's largest industry (Ross, 2012, p. 3), and given its vast vested interests in business as usual, the role of the fossil fuel sector is worthy of full attention in another study.

1.4. Societal relevance: The climate change problem

The climate change problem refers to the cumulative emissions of GHGs resulting from the energy, food and other development sectors that cause global warming. The societal dimension is on the impact side but also on the emission side. The impact side refers to the impacts of climate change—which include rising temperatures, melting glaciers, rising sea levels, changed precipitation patterns, ocean acidification and potentially irreversible changes such as impacts on ocean circulation patterns and polar melting (IPCC, 2014, 2019, 2021). The global temperatures have already risen by 1.2 °C over pre-industrial levels. If they rise beyond 2 °C, they will reach a level that has never been experienced in the Holocene and could have major impacts on society. These are levels that bring us into a no-analogue zone—where we have absolutely no idea how and if we can survive as a human species or what the impacts will be on other ecosystems.

Climate change has already advanced to such an extent that it is seen as a driver of all other ecosystem challenges—whether in relation to oceans, land, fresh water and biodiversity (Ekins et al., 2019). It thereby impacts on the lives and livelihoods of billions of people who depend directly on nature for their survival (i.e. about 70% of the world's poor depend directly on nature). It will and does disproportionately affect developing countries and poor and vulnerable people (Gupta et al., 2020). It is impacting on water systems, food systems, housing and infrastructure. It is already leading to extreme weather events, including droughts and disasters, and such events affect millions annually (Ekins et al., 2019). It is no wonder that the European Parliament now sees climate change as an "emergency" (European Parliament, 2019). This clearly means that action needs to be taken urgently and yet democratically to address the problem.

This brings us to the issue of mitigation: the need to mitigate the emissions of GHGs. The 1.5–2 °C objective requires transformative changes in the way societies develop and produce (IPCC, 2014), especially massive reductions in their GHGs (IPCC, 2013, p. 19). This implies decarbonising electricity generation and energy supplies in the industry and transport sector (IPCC, 2014). This is because the bulk of GHGs comes from the use of fossil fuel (IPCC, 2014). There is no alternative to phasing out fossil fuels. Some argue that bio-energy with carbon capture and storage (BECCS) technology can help us buy time. These options have been included in scenarios prepared by or for the Intergovernmental Panel on Climate Change (IPCC). However, there is growing evidence that bio-energy at the scale required is not possible (because we do not have the land and other resources to grow the required amount of bio-fuels) and carbon capture and storage is very expensive and risky (Ekins et al., 2019).

This brings us back to the issue of (equitably) replacing fossil fuels with energy sources that are clean and do not emit GHGs. Clearly this also requires us to rethink the continuing and growing demand for energy. Phasing out fossil fuels brings with it some major challenges. First, fossil fuels are asymmetrically distributed globally and therefore embody spatial implications. Second, fossil fuel production and consumption are the driver of most economies worldwide and phasing out fossil fuels is seen as a major risk to economies (see Chapters 4 to 7). Third, since many sectors of the economy including households are using fossil fuels—changing the system leads to huge costs to everyone and this also leads to social movements resisting policies to phase out fossil fuels or make them more expensive as was the case when the public protested the rising price of fuels in France. At the same time there are other social movements that object to the way the extractive industries function and to climate change. Unpacking this narrative is what this book aims to do.

1.5. Policy relevance: The climate change regime and Agenda 2030

1.5.1. The climate change regime and fossil fuels

This book is policy relevant at the international level as it aims to support the implementation of the climate change regime (for details of its history, see Gupta, 2014) and Agenda 2030. The two regimes are briefly explained below as well as their implications for this book. (It is also policy relevant at country level, as the empirical chapters show.)

We believe that the history of the negotiations is critical to understanding the temporal dimension of the politics, economics and environmental aspects of the problem. With the General Assembly adopting an intergovernmental negotiating process on climate change in 1989, negotiations on the United Nations Framework Convention on Climate Change (UNFCCC, 1992) began leading to its adoption in 1992. This convention specified the urgency of the problem, recognised the differentiated responsibilities and respective capabilities of rich and poor countries and outlined a list of targets, policies and measures for all countries. It proposed a financial mechanism where industrialised countries would provide "new and additional" resources to enable developing countries to address mitigation and adaptation challenges and thereby promote the implementation of the convention. It also proposed market mechanisms. The convention recognised that rich countries would have to reduce their GHGs emissions to make space for the legitimate increase in the emissions of developing country. However, there was also discussion about whether the oil-rich developing countries should be compensated for the loss of revenue they might face if they were to phase out their emissions and this resulted in the inclusion of Article 8.[2] In fact, over the next 10 years, this debate on whether industrialised countries should compensate members of OPEC (Organization of the Petroleum Exporting Countries) for their loss of income (Gonn, 2009) hampered other discussions on the financial mechanism. This might provide some background to new demands to compensate developing countries to leave their fossil fuels underground (see Chapter 4).

In 1997, a follow-up protocol—the Kyoto Protocol to the United Nations Framework Convention on Climate Change—was adopted (Kyoto Protocol, 1997). It specified GHG emissions targets for the industrialised world for the period from 2008 to 2012, which would have, if achieved collectively, amounted to a 5.2% reduction in relation to 1990 levels. However, the US and Canada did not participate. The US claimed that the protocol was "fatally flawed" as it exempted developing countries from quantitative targets (White House, 2001, cited in Byrne et al., 2007; Harris, 2009). Moreover, the achievement of the targets of other industrialised countries was facilitated by the use of market mechanisms (e.g. the Clean Development Mechanism, Joint Implementation, Emissions Trading) and so this did not really cause a major dent in fossil fuel use in these countries or in the developing world (e.g. Sinn, 2012). While the market mechanisms were meant to unleash a major process of decarbonisation worldwide, this did not eventually occur, not least because of the non-participation of North America, but also because the project-based approach did not lead to programmatic change. Furthermore, although the protocol indicated that the industrialised countries would provide "new and additional" financial resources to enable developing countries to implement the protocol, which was meant to be over and above the 0.7% of gross domestic product (GDP) that was being used for development assistance, industrialised countries interpreted this as over and above the actual development cooperation resources which was significantly below 0.7% (Gupta & Van der Grijp, 2010; Stadelmann et al., 2013, p. 1; Josephson, 2017). Such new and additional finances remain a hot issue even today as this book demonstrates. In 2012, the Doha Amendment to the Kyoto Protocol was agreed; it aimed at identifying targets for as many countries as possible to continue the momentum of the Kyoto Protocol for the period from 2012 to 2020. Despite the fact that Barack Obama was at

2 "[T]he Parties shall give full consideration to what actions are necessary under the Convention, including actions related to funding, insurance, and the transfer of technology, to meet the specific needs and concerns of developing countries arising from … the impact of the implementation of response measures, especially on:… Countries whose economies are highly dependent on income generated from the production, processing and export, and/or on consumption of fossil fuels and associated energy-intensive products."

that time president of the US and despite the large-scale ratification of the amendment by developing countries, the Doha Amendment only entered into force on the last day of 2020—and much of the momentum was lost.

It is then logical that the Paris Agreement (2015) was greeted with so much acclaim. It tried to re-energise the legal momentum in the climate change-negotiating process. Finally, after 23 years, it identified an overarching objective, thereby articulating clearly in quantitative terms the long-term objective of the agreement and the need for financial coherence in Article 2.[3] This book will focus on these two issues because they provide the international context for phasing out fossil fuel use and because they highlight the need for financial coherence. As mentioned above, the Paris Agreement does not mention fossil fuels or establish any targets relating to fossil fuels. However, achieving the long-term objective is only possible by phasing out net fossil fuels by 2050 (see 1.5.1).

The Paris Agreement set up a process for implementation and monitoring of the implementation. But it gave up the idea that the industrialised countries should lead first and make space for developing countries; rather, *all* countries were required to identify targets for themselves in nationally determined contributions (NDCs)—thus moving away from an equitable allocation to a post-equity narrative where everyone does what they want to and can, and the equity narrative shifted from common but differentiated responsibilities to respective capabilities in the light of different national circumstances; and assistance from industrialised countries to developing countries is now more or less equivalent to "capacity building" and loans within the Green Climate Fund. The lofty ideals that had brought 193 countries together to adopt the UNFCCC in 1992 and ratify it rapidly by 1994 have been watered down over the years. For example, calls for self-determined NDCs as a non-legally binding component of the Paris Agreement is problematic; this concession was made to enable President

3 "This Agreement, in enhancing the implementation of the Convention, including its objective, aims to strengthen the global response to the threat of climate change, in the context of sustainable development and efforts to eradicate poverty, including by: (a) Holding the increase in the global average temperature to well below 2 °C above pre-industrial levels and pursuing efforts to limit the temperature increase to 1.5 °C above pre-industrial levels, recognising that this would significantly reduce the risks and impacts of climate change; (b) Increasing the ability to adapt to the adverse impacts of climate change and foster climate resilience and low greenhouse gas emissions development, in a manner that does not threaten food production; and (c) Making finance flows consistent with a pathway towards low greenhouse gas emissions and climate-resilient development. This Agreement will be implemented to reflect equity and the principle of common but differentiated responsibilities and respective capabilities, in the light of different national circumstances."

Obama to bypass the US Senate and ratify the agreement. However, bypassing the Senate was also possible for President Trump, who then withdrew from the Paris Agreement (effective from 4 November 2020). Member of the House of Representatives Alexandria Ocasio-Cortez proposed a Green New Deal though it was met with Trump's opposition. And shortly after his inauguration, President Biden again bypassed the Senate to rejoin the Paris Agreement (20 January 2021). But the bottom line is that the way the Paris Agreement is designed does not guarantee US participation and, worse still, by excluding the national commitments and the financial promise of an extra $100 billion annually to finance developing countries from the legally binding part of the Paris Agreement text, it has watered down the commitments even further. This has also meant that the $100 billion to be provided annually to address climate change is not yet on the table even though the European Parliament has urged the European Union (EU) to double its contributions to the Green Climate Fund (European Parliament, 2019). There is also increasing evidence that development assistance money is being re-labelled as climate money.

The first set of NDCs under the Paris Agreement aim to reduce global emissions by 40% by 2030 in relation to 1990 levels (Paris Agreement, 2015). Some NDCs include bans and moratoriums on specific types of fossil fuels (e.g. China, India, and Spain focus on coal) and some focus on specific locations (e.g. the US moratorium on "oil and gas exploration in some areas of the Arctic and Atlantic" [SEI et al., 2019, p. 43]). The NDCs include conditional and unconditional commitments and studies show that these reductions are likely to be 50% higher than what is necessary to limit average global temperatures by 2 °C, and 120% higher than what is needed for a 1.5 °C rise (SEI et al., 2019, p. 2; WRI, 2019). In fact, the ambitions in the NDCs have to be ratcheted up by five times to achieve the 1.5 °C goal (UNEP, 2019; King & Van den Bergh, 2019) and this does not even address the fact that many countries are falling short in achieving their NDC GHG pledges (Aldy et al., 2017; Roelfsema et al., 2020). It should be noted that shipping and aviation emissions as well as those from defence activities are not explicitly included in this.

At the 26th Conference of the Parties of the UNFCCC in November 2021 (COP26), for the first time, an express objective towards achieving net zero emissions and the goal of 1.5 °C, by accelerating the phase-out of fossil fuels, in particular coal, was articulated (Carver, 2022), correcting the omission to mention fossil fuels explicitly in the Paris Agreement (Harvey, 2021). It called for the "phase down of unabated coal power and phase out of inefficient fossil fuel subsidies, while providing targeted support to the poorest and most vulnerable in line with national circumstances and recognising the need for support towards a just transition" (UNFCCC, 2021). A total of 190 countries agreed to phase down coal power, which could decrease new coal power plants by 76%, and over 40 countries supported the Global Coal to Clean Power Transition Statement (Carver, 2022). They also agreed to reassess and strengthen their NDCs up to 2030 (Danaher, 2021). The newer submissions could lead to a combined increase in GHG emissions of 13.7% in 2030 above 2010 (Coleman, 2021), while a 1.5 °C target would require reducing emissions by 45% by 2030 (Coleman, 2021).

A critical reflection of the climate change regime shows that it has taken a symptomatic approach to addressing climate change focusing primarily on emissions, but scarcely questioning the production, distribution and consumption patterns or the underlying neoliberal capitalist paradigm that is responsible for the current rapacious growth model. Its focus on technology transfer, market mechanisms and financial mechanisms tends to reproduce Western lifestyles in the Global South rather than questioning its fundamental premises (Gupta, 2014; Chapter 2). This is a point we return to in the last chapter.

1.5.2. The 2030 Agenda and fossil fuels

This brings us to a discussion of the global 2030 Agenda and fossil fuels. This agenda has been initiated by member states of the United Nations and consists of a long list of development targets that are headed under 17 Sustainable Development Goals (SDGs), which have been internationally adopted (UNGA, 2015). This successor of the Millennium Development Goals has been praised for its transformation potential and critiqued for not having sufficient ambition and "teeth" and insufficiently addressing the fundamental sources of inequality in the international political and economic system (Bond, 2018). Despite these reservations, from the perspective of this book, Agenda 2030 is important for four reasons. First of all, for the first time in human history, social goals such as poverty eradication have been put on par with ecological and economic goals at the global level—a remarkable achievement. We have had treaties on trade, investment and the environment, but we have never had treaties on social issues such as poverty reduction, energy access and so on. Agenda 2030 corrected this by outlining 17 goals for the global community and clearly prioritising poverty reduction.[4] In addition to poverty, the goals of Agenda 2030 emphasise access to water,

4 "We recognise that eradicating poverty in all its forms and dimensions, including extreme poverty, is the greatest global challenge and an indispensable requirement for sustainable development" (UNGA, 2015, para. 2).

food, energy, infrastructure, housing, education, health and so on, and it focuses on reducing inequality. It thus articulates a commitment to justice at a global level, which goes beyond a conservative commitment to a human rights-based approach. However, the agenda fails on corrective justice (Gupta & Schmeier, 2020). Agenda 2030 is strongly embedded in human rights law and in the Rio Declaration on Environment and Development (UNCED, 1992).

Second, the 2030 Agenda not only prioritises poverty as the greatest challenge, but it also emphasises that all goals have to be treated in an integrated and indivisible manner.[5] This is remarkable because it basically argues that achieving one goal at the cost of other goals is not possible, that synergies have to be sought between different goals; something this book tries to assess. Third, the SDGs require that *all* countries become sustainable, seeing also industrialised countries as unsustainable (Hajer et al., 2015). The SDGs require significant change in consumption patterns and reduction in resource use in the Global North (Leal Filho et al., 2019). There is also a call for participation by the private sector (Scheyvens et al., 2016) and other non-state actors. Fourth, the 2030 Agenda is the result of breakthroughs in international negotiation requiring countries from the Global North and South to share one of 30 "seats" and to come up with a common position before negotiating internationally (Chasek & Wagner, 2016). It is the world's largest crowdsourcing event where 7 million people, especially those in the developing world, were consulted (through fax, interviews, telephone calls, etc.) with 58% of the respondents aged between 16 and 30 (Gellers, 2016).

Clearly the document is the result of negotiations and hence embodies some incoherence within it, which results from the compromises reached. Two key issues here are: (a) its focus on "partnership" and collaboration while calling for "full permanent sovereignty," which developing countries wanted as a way to protect their rights over their resources and policies (Gupta & Ceylan, 2020); and (b) its focus on "growth" (mentioned 16 times), reflecting the wishes of those who see GDP growth as an imperative while using the adjective "inclusive" (41 times), reflecting the wishes of critical thinkers (Gupta & Vegelin, 2016). The compromise was that "growth" is almost always prefixed by "inclusive" and is often in relation to small-scale enterprises or least developed countries.

Two of the 2030 Agenda's 17 goals are central to this book: the climate change goal (#13) and the energy goal (#7). The climate target aims to "[i]ntegrate climate change measures into national policies, strategies and

5 "We are announcing today 17 Sustainable Development Goals with 169 associated targets which are integrated and indivisible" (UNGA, 2015, para. 18).

planning" (Target 13.2) and "[i]mprove education, awareness-raising and human and institutional capacity on climate change mitigation, adaptation, impact reduction and early warning" (Target 13.3). It requires a joint effort by both the public and the private sectors (Target 13.2) and enables reducing impact (Target 13.3) and calls for generating $100 billion annually (Target 13.A). However, the climate change goal is deliberately vague, deferring to the UNFCCC regime. (It was included, because the negotiators wanted to ensure that climate change was seen as a top global priority.) The energy goal focuses on universal access to affordable, reliable and modern energy services and calls for increasing the share of renewables in the global energy mix. However, it also calls for "cleaner fossil fuel technology" (Target 7.A).

Access to energy is linked to reducing poverty (#1), reducing hunger (#2), enhancing health (#3), the water goal (#6), sustained, inclusive and sustainable economic growth and employment (#8), resilient infrastructure (#9), reduced inequality (#10), sustainable cities (#11), sustainable production and consumption patterns (#12), sustainable use of land (#15), and peace (#16). This implies that in analysing fossil fuel phase-out, we need to consider a number of related issues: (a) socio-ecological issues—(i) access to income/jobs; (ii) access to services of food, water, land, energy, infrastructure, housing; (iii) access to nature's contributions to people; and (iv) procedural rights of non-discrimination, information, decision-making, recognition, courts (Gupta & Lebel, 2020); (b) economic issues—impacts on the economy, infrastructure; (c) broader justice issues—impacts on inequality, production and consumption patterns, and peace; and (d) in relation to North–South issues—impacts on least developed countries, small island states and the right to development (see Table 1.1).

Table 1.1. How two of the Agenda 2030 SDGs link to the phasing out of fossil fuel

Central SDGs	Issue clusters	Synergies & trade-offs
Climate change (Goal 13): Implicitly phase out fossil fuel; Energy (Goal 7): Promote affordable, reliable modern energy; increase renewables and clean fossil fuels	Socio-ecological access	Jobs/income (energy related) Services (affordable, reliable, modern energy) NCPs (energy extraction, production, use and impacts on nature) Procedural rights (rights related to energy)
	Socio-ecological allocation	Inequality (energy-related inequalities) Production/consumption patterns (and use of energy)
	North–South	Issues of developing countries vis-à-vis energy Right to development and energy
	Economic	Economy (link between energy and economy) Infrastructure (energy related)

This discussion has two implications for our analysis in this book: (a) Any analysis of the issue of LFFU needs to take the other goals into account (see Table 1.1); and (b) the issue of dividing responsibilities between states on fossil fuels remains contested as states are seen to have full sovereignty over their fossil fuel resources. Despite calls for global solidarity, the 2030 Agenda does not call for liability or responsibility from those who caused harm to others.

1.6. Focus on Africa and Latin America

While this book takes a global perspective it pays special attention to Africa and Latin America, with case studies on South Africa and Ecuador. This focus is valuable because the negative current and future effects of fossil fuels particularly harm vulnerable groups in low- and middle-income countries, such as Indigenous peoples, peasant communities, poor families and women. By doing so, the extraction and use of fossil fuels is a major source of socio-environmental and climate injustice (cf. Bond, 2016; Pearse, 2016; Vásquez, 2014), and this has sparked various LFFU initiatives in these regions. However, as mentioned above, leaving fossil fuels underground also involves major trade-offs for different actors in such countries, ranging from very vulnerable to very powerful groups. This complexity is especially clear in countries with large fossil fuel reserves. While the empirical findings and analysis of the case studies on South Africa and Ecuador evidently cannot be generalised to Africa, Latin America and the Global South as a whole, they offer important insights and pointers to our understanding of the global dynamics of LFFU, including North–South interactions.

1.6.1. South Africa

Africa has an estimated 126 billion barrels (Gbbl) of proven oil reserves and 647 trillion cubic feet (Tcf) of proven natural gas reserves, equivalent to roughly 7.5% and 9% of global reserves, respectively. Across the continent, almost 85% of produced crude oil and petroleum products and 41% of natural gas is exported. Coal is less ubiquitous but undoubtedly relevant, with South Africa, Zimbabwe, Botswana, Mozambique, Tanzania and Nigeria's proven reserves combining to 14.8 billion tonnes of coal, or roughly 1.4% of global reserves (computed using data from BP, 2020).

Within Africa, South Africa is a key case to consider in the LFFU narrative as it is home to one of the world's most carbon-intensive economies.

Coal continues to be the mainstay of the South African energy system, supplying as much as 70% of installed power-generation capacity (IEA, 2021). Simultaneously, South Africa is one of the world's most unequal societies, with a national unemployment rate of 30–35%. South Africa's coal sector generates an estimated 90,000 direct jobs (e.g. mining) and 170,000 indirect jobs (e.g. equipment manufacturing) while also being responsible for considerable local pollution (Res4Africa, 2021).

Phasing out fossil fuels in the South African context calls for revamping the minerals–energy complex (MEC) (Fine & Rustomjee, 1996), a long-standing legacy from the apartheid era that has evolved into the existing, carbon-dependent national regime. Restructuring the MEC into a climate-friendly configuration faces a multitude of challenges that extend far beyond those of employment—as was just alluded to—including but not limited to: energy access, health and safety, international trade, racism, sexism and gender-based violence, and colonial and imperial legacies.

South Africa's proposed modification through the NDCs have been labelled as highly insufficient, with both policies and actions and fair share targets estimated to lead to warming over 2 °C and up to 3 °C (Climate Action Tracker, 2022). One contributing factor to this misalignment stems from South Africa's plan to remain 51% coal dependent in installed power capacity by 2030 (DMRE, 2019). South Africa's government is clearly focusing on economic growth and does not appear to take alternative development paths into account. South Africa's MEC and NDCs and the responses of social movements and their intricacies are further explored in Chapters 5 and 6.

1.6.2. Ecuador

Latin America has 337.7 billion barrels of proven oil reserves and 7,954 billion cubic metres of proven natural gas reserves. This is equivalent to 19% and 4% of global reserves, respectively. Venezuela alone holds 17% of oil reserves and 3% of natural gas reserves worldwide. The region's coal proven reserves account for 1.5% with 16,118 Mt with Brazil and Colombia as the main producers (OLADE, 2021). Moreover, 40% of the world's unconventional oil reserves are in Latin America (Valdivia & Lyall, 2018, p. 464). In 2020, the region's CO_2 emissions amounted to 1,435 Mt, equivalent to 4.5% of global emissions (OLADE, 2021).

Ecuador is among Latin America's largest oil exporters. After Venezuela and Brazil, it has the region's third-largest oil reserves (8.3 billion barrels) and is the fifth-largest oil producer in the region, producing 534,000 b/d in 2019.

In 2020, 70% of its oil production was for export, accounting for almost half of its export earnings and a fifth of public sector revenues (EIA, 2021). Oil is the main source of primary energy consumption (around 75%), but since 2010 Ecuador's hydroelectric capacity has surpassed the national demand and has become progressively the main source of electricity production (CELEC, 2021). Still, Ecuador's economy remains highly dependent on oil exports and oil extraction continues despite being the cause of several socio-environmental conflicts. The arrival of the oil industry in the 1970s to the Amazon—where its oil reserves are located—meant an abrupt incursion into the territories of many Indigenous nationalities (Quintero & Silva, 1991; Sawyer, 2004).

Oil extraction in Ecuador is a deeply politicised matter, which for years has triggered the contestation of civil society. On the one hand, the experiences of oil exploitation include human rights violations related to the colonisation of Amazon territories and the pollution of livelihood means as oil spills pollute water and soils, and ever-burning torches taint the air. The Amazon populations close to the extraction sites are among the poorest and have the highest cancer rates in the country. The lawsuit that 30,000 Amazon Indigenous people and peasants filed against Texaco, now ChevronTexaco, for the impacts generated during its operations is an iconic case of Ecuador's history of social organisation vis-à-vis the oil industry (Beristain et al., 2009). On the other hand, Ecuador's economy still depends on the revenue generated by oil exports, making its economy vulnerable to price fluctuations (Andrade, 2016; Fontaine, 2007). Moreover, its fossil fuel subsidies are among the highest in Latin America, and their elimination is a matter of contention. Even when the subsidy has regressive distributional effects, it plays a politically symbolic role on the imaginary of revenue redistribution in the energy and production sectors. Moreover, the precarious job opportunities and compensations of the oil industry has led to clientelism for the Amazon populations who live amid unemployment and poverty.

Ecuador's society has pioneered important organisational efforts to contest fossil fuels. Next to the above-mentioned trial against Texaco—in 2011 a court ruled that the company should pay $9.5 billion (Lu et al., 2017), but the company has contested the ruling ever since—Ecuadorian civil society groups were among the first to call for leaving fossil fuels underground in international fora, as part of the organisation Oilwatch. These calls became the seed to what later took the form of the Yasuní–ITT Initiative launched by the Ecuadorian government in 2007 to prevent oil extraction from Yasuní National Park in exchange for the compensation from industrialised countries amounting to half the revenues Ecuador would otherwise receive from extraction (see Chapter 8). The constitutional enactment of the rights of nature (Espinosa, 2019) and critiques of the model of development under the concepts of *buen vivir* are part of wider ongoing debates among social environmental movements and scholarship in Latin America and beyond (Radcliffe, 2012; Svampa, 2019). Chapter 4 examines the way civil society actors in Latin America and in Ecuador continue to mobilise ideas to phase out fossil fuel extraction, as well as the challenges and opportunities to achieve this goal.

1.7. Transdisciplinary methodology

This research is based on a transdisciplinary methodology. It combines different academic disciplines and approaches—namely law, politics, policy sciences, anthropology, sociology and engineering. It merges these different disciplines in the identification and analysis of actors, arguments and approaches. Moreover, our understanding of LFFU actors, arguments and approaches has been informed by in-person and online meetings and discussions about our research findings with a variety of practitioners—activists, governmental and NGO experts—from different countries and generations.

In terms of methodological steps, this research is based on (a) a combination of rapid and systematic literature reviews on a number of theoretical and conceptual issues (see Chapter 2). It has used (b) systematic content analysis to assess the laws and policies of different actors—at the intergovernmental level, at the state level, at the level of different actors and at the level of non-state actors and civil society. Social movements are rather varied and flexible, and it can be laborious to assess their main arguments and approaches (for instance, requiring interviews). Non-state actors (such as NGOs) are sometimes easier to analyse, more willing to be interviewed and their extensive documentation enables content analysis. Hence, we have also analysed news clippings and other sources of information (checking for legitimacy in order to screen out fake news) to assess the role and impact of social movements. We have then used (c) quantitative data analysis of relevant data collected from a range of different sources. Finally, we have focused on (d) two major case studies (of Ecuador and South Africa). The single case study approach in the two continents is embedded in an understanding of regional dynamics and we hope to be able to draw lessons from each case study that are relevant for other similar countries in the same or different continents.

1.8. Structure of this book

This book is structured as follows. Chapter 2 reviews the theoretical and conceptual strands that have been used in the process of the research work. It brings these strands together in a coherent conceptual framework. Chapter 3 examines the role of social movements worldwide and the way in which they have used direct action, including the courts, to stop fossil fuels or promote renewables. Chapter 4 looks at social movements in Latin America with a special focus on Ecuador. Chapter 5 examines the arguments of different actors with respect to LFFU in South Africa. Chapter 6 explores the social movements in South Africa. Chapter 7 assesses the role of investors in the fossil fuel sector with a special focus on aid and export credit agencies, pension funds, banks and philanthropy. Chapter 8 looks at supply-side measures and Chapter 9 concludes the book by bringing the different storylines together.

References

Aengenheyster, M., Feng, Q. Y., Van der Ploeg, F., & Dijkstra, H. A. (2018). The point of no return for climate action: Effects of climate uncertainty and risk tolerance. *Earth System Dynamics, 9*(3), 1085–1095. https://doi.org/10.5194/esd-9-1085-2018

Aldy, J. E., Pizer, W. A., & Akimoto, K. (2017). Comparing emissions mitigation efforts across countries. *Climate Policy, 17*(4), 501–515. https://doi.org/10.1080/14693062.2015.1119098

Andrade, P. A. (Ed.). (2016). *La gobernanza ambiental en Ecuador. Historia, presente y desafíos*. Universidad Andina Simón Bolívar/Corporación Editora Nacional.

Beristain, C. M., Páez Rovira, D., & Fernández, I. (2009). *Las palabras de la selva: Estudio psicosocial del impacto de las explotaciones petroleras de Texaco en las comunidades amazónicas de Ecuador.* Hegoa.

Bond, P. (2016, Winter). Who wins from "climate apartheid"? African climate justice narratives about the Paris COP21. *New Politics*, 122–129.

Bond, P. (2018). Climate debt, community resistance and conservation alliances against Kwazulu-Natal coal mining at Africa's oldest nature reserve. In B. Engels & K. Dietz (Eds.), *Climate change in Africa: Social and political impacts, conflicts, and strategies* (pp. 314–337). Peter Lang.

Bos, K., & Gupta, J. (2018). Climate change: The risks of stranded fossil fuel assets and resources to the developing world. *Third World Quarterly, 39*(3), 436–453. https://doi.org/10.1080/01436597.2017.1387477

Bos, K., & Gupta, J. (2019). Stranded assets and stranded resources: Implications for climate change mitigation and global sustainable development. *Energy Research & Social Science, 56*. https://doi.org/10.1016/j.erss.2019.05.025

BP. (2020). *BP Sustainability Report 2020*. https://www.bp.com/content/dam/bp/business-sites/en/global/corporate/pdfs/sustainability/group-reports/bp-sustainability-report-2020.pdf

Byrne, J., Hughes, K., Rickerson, W., & Kurdgelashvili, L. (2007). American policy conflict in the greenhouse: Divergent trends in federal, regional, state, and local green energy and climate change policy. *Energy Policy, 35*(9), 4555–4573.

Carbon Tracker Initiative. (2019, November 1). Oil majors must cut output by average of 35% to 2040 to hit Paris targets: ConocoPhillips and Exxon must cut most, Shell Least [Press release]. https://carbontracker.org/balancing-the-budget-press-release/

Carbon Tracker Initiative. (N.d.). Terms list. https://carbontracker.org/resources/terms-list/#carbon-budgets

Carver, D. (2022). What were the outcomes of COP26? House of Commons Library. https://commonslibrary.parliament.uk/what-were-the-outcomes-of-cop26/

CELEC [Corporación Eléctrica del Ecuador]. (2021). CELEC EP genera y transmite más del 90 por ciento de la energía eléctrica limpia que consume el país y exporta a los países vecinos. https://www.celec.gob.ec/gensur/index.php/435-celec-ep-genera-y-transmite-mas-del-90-por-ciento-de-la-energia-electrica-limpia-que-consume-el-pais-y-exporta-a-los-paises-vecinos

Chasek, P. S., & Wagner, L. M. (2016). Breaking the mold: A new type of multilateral sustainable development negotiation. *International Environmental Agreements: Politics, Law and Economics, 16*, 397–413. https://doi.org/10.1007/s10784-016-9320-2

Clark, P. (2020, October 9). Yellen and Carney sound alarm over stranded assets. *Financial Times*. https://www.ft.com/content/d94e4bae-4ce6-4db2-9755-2d976de2892f

Climate Action Tracker. (2022, October 28). South Africa country summary. https://climateactiontracker.org/countries/south-africa/

Coleman, C. (2021). COP26: Aims, goals and progress: Library briefing. UK Parliament House of Lords Library. https://researchbriefings.files.parliament.uk/documents/LLN-2021-0031/LLN-2021-0031.pdf

Danaher, L. (2021). COP26 keeps 1.5C alive and finalises Paris Agreement: UN Climate Change Conference (COP26) at the SEC—Glasgow 2021. https://ukcop26.org/cop26-keeps-1-5c-alive-and-finalises-paris-agreement/

DMRE [Department of Mineral Resources and Energy]. (2019). Integrated Resource Plan. http://www.energy.gov.za/IRP/2019/IRP-2019.pdf

EIA [US Energy Information Administration]. (2021). Country analysis executive summary: Ecuador. https://www.eia.gov/international/content/analysis/countries_long/Ecuador/ecuador.pdf

EIB [European Investment Bank]. (2019). EU Bank launches ambitious new climate strategy and energy lending policy. https://www.eib.org/en/press/all/2019-313-eu-bank-launches-ambitious-new-climate-strategy-and-energy-lending-policy

Ekins, P., Gupta, J., & Boileau, P. (2019). *Global environment outlook–GEO-6: Healthy planet, healthy people.* Cambridge University Press.

Espinosa, C. (2019). Interpretive affinities: The constitutionalization of rights of nature, *Pacha Mama*, in Ecuador. *Journal of Environmental Policy & Planning, 21*(5), 608–622. https://doi.org/10.1080/1523908X.2015.1116379

European Parliament. (2019). Resolution of 28 November 2019 on the Climate and Environment Emergency (2019/2930(RSP)). https://www.europarl.europa.eu/doceo/document/TA-9-2019-0078_EN.html

Fine, B., & Rustomjee, Z. (1996). *The political economy of South Africa: From minerals–energy complex to industrialisation.* Routledge.

Fontaine, G. (2007). Verde y negro: Ecologismo y conflictos por petróleo en el Ecuador. In G. Fontaine, G. Van Vliet & R. Pasquis (Eds.), *Polıticas ambientales y gobernabilidad en America Latina* (pp. 223–256). FLACSO.

Geddes, A., et al. (2020, November 9). *Doubling back and doubling down: G20 scorecard on fossil fuel funding.* International Institute for Sustainable Development. https://www.iisd.org/system/files/2020-11/g20-scorecard-report.pdf

Gellers, J. C. (2016). Crowdsourcing global governance: Sustainable Development Goals, civil society, and the pursuit of democratic legitimacy. *International Environmental Agreements: Politics, Law and Economics, 16*(3), 415–432. https://doi.org/10.1007/s10784-016-9322-0

Gonn, A. (2009, November 23). OPEC head backs Saudi compensation claims over lost oil revenues. *Jerusalem Post.* https://www.jpost.com/business/opec-head-backs-saudi-compensation-claims-over-lost-oil-revenues

Gupta, J. (1997). *The Climate Change Convention and developing countries: From conflict to consensus?* Kluwer.

Gupta, J. (2014). *The history of global climate governance.* Cambridge University Press.

Gupta, J., Rempel, A., & Verrest, H. (2020). Access and allocation: The role of large shareholders and investors in leaving fossil fuels underground. *International Environmental Agreements: Politics, Law and Economics, 20*(2), 303–322. https://doi.org/10.1007/s10784-020-09478-4

Gupta, J., & Ceylan, U. (2020). The global environment outlook and its implications for international law: Is law increasingly falling behind? In G. Z. Capaldo (Ed.), *The global community yearbook of international law and jurisprudence* (pp. 49–82). Oxford University Press.

Gupta, J., & Lebel, L. (2010). Access and allocation in earth system governance: Water and climate change compared. *International Environmental Agreements: Politics, Law and Economics, 10*(4), 377–395. https://doi.org/10.1007/s10784-010-9139-1

Gupta, J., & Lebel, L. (2020). Access and allocation in earth system governance: Lessons learnt in the context of the Sustainable Development Goals. *International Environmental Agreements: Politics, Law and Economics,* 20, 393–410. https://doi.org/10.1007/s10784-020-09486-4

Gupta, J., & Schmeier, S. (2020). Future proofing the principle of no significant harm. *International Environmental Agreements: Politics, Law and Economics, 20*(4),731–747. https://doi.org/10.1007/s10784-020-09515-2

Gupta, J., & Van der Grijp, N. (2010). *Mainstreaming climate change in development cooperation: Theory, practice and implications for the European Union.* Cambridge University Press.

Gupta, J., & Vegelin, C. (2016). Sustainable Development Goals and inclusive development. *International Environmental Agreements: Politics, Law and Economics, 16*(3), 433–448. https://doi.org/10.1007/s10784-016-9323-z

Hajer, M., et al. (2015). Beyond cockpit-ism: Four insights to enhance the transformative potential of the Sustainable Development Goals. *Sustainability, 7*(2), 1651–1660. https://doi.org/10.3390/su7021651

Harris, P. G. (2009). Beyond Bush: Environmental politics and prospects for US climate policy. *Energy Policy, 37*(3), 966–971. https://doi.org/10.1016/j.enpol.2008.10.042

Harvey, F. (2021, November 10). COP26 draft text annotated: What it says and what it means. *The Guardian.* https://www.theguardian.com/environment/ng-interactive/2021/nov/10/cop26-draft-text-annotated-what-it-says-and-what-it-means

Hicks, R. L., Parks, B. C., Roberts, J. T., & Tierney, M. J. (2008). *Greening aid? Understanding the environmental impact of development assistance.* Oxford University Press.

IEA [International Energy Agency]. (2021). South Africa. https://www.iea.org/countries/south-africa

IPCC [Intergovernmental Panel on Climate Change]. (2013). Summary for policymakers. In *Climate change 2013: The physical science basis. Contribution of Working Group I to the fifth assessment report of the of the Intergovernmental Panel on Climate Change.* [T. F. Stocker, D. Qin, G.-K. Plattner, M. Tignor, S. K. Allen, J. Boschung, A. Nauels, Y. Xia, V. Bex & P. M. Midgley (Eds.)]. Cambridge University Press. https://www.ipcc.ch/site/assets/uploads/2018/03/WG1AR5_SummaryVolume_FINAL.pdf

IPCC [Intergovernmental Panel on Climate Change]. (2014). *Climate change 2014: Synthesis report. Contribution of Working Groups I, II and III to the fifth assessment report of the Intergovernmental Panel on Climate Change.* [Core Writing Team, R. K. Pachauri & L. A. Meyers (Eds.)].

IPCC [Intergovernmental Panel on Climate Change]. (2019). *Global warming of 1.5 °C.* https://www.ipcc.ch/sr15/

IPCC [Intergovernmental Panel on Climate Change]. (2021). *AR6 synthesis report: Climate change 2022.* https://www.ipcc.ch/report/sixth-assessment-report-cycle/

Josephson, P. (2017). *Common but differentiated responsibilities in the climate change regime: Historic evaluation and future outlooks.* PhD thesis, Stockholm University. https://www.diva-portal.org/smash/get/diva2:1134510/FULLTEXT01.pdf

King, L. C., & Van Den Bergh, J. C. J. M. (2019). Normalisation of Paris Agreement NDCs to enhance transparency and ambition. *Environmental Research Letters, 14*(8). https://doi.org/10.1088/1748-9326/ab1146

Kyoto Protocol. (1997). Kyoto Protocol to the United Nations Framework Convention on Climate Change (adopted on 11 December 1997, entered into force 16 February 2005), 2303 UNTS 192.

Leal Filho, W., Tripathi, S. K., Andrade Guerra, J. B. S. O. D., Giné-Garriga, R., Orlovic Lovren, V., & Willats, J. (2019). Using the Sustainable Development Goals towards a better understanding of sustainability challenges. *International Journal of Sustainable Development & World Ecology, 26*(2), 179–190. https://doi.org/10.1080/13504509.2018.1505674

Linquiti, P., & Cogswell, N. (2016). The carbon ask: Effects of climate policy on the value of fossil fuel resources and the implications for technological innovation. *Journal of Environmental Studies and Sciences, 6*(4), 662–676. https://doi.org/10.1007/s13412-016-0397-2

Lu, F., Valdivia, G., & Silva, N. L. (Eds.). (2017). *Oil, revolution, and Indigenous citizenship in Ecuadorian Amazonia.* Palgrave Macmillan.

McDermott Hughes, D. (2017). *Energy without conscience: Oil, climate change, and complicity.* Duke University Press.

Mercure, J., et al. (2018, July). Macroeconomic impact of stranded fossil fuel assets. *Nature Climate Change, 8*, 588–593. https://doi.org/10.1038/s41558-018-0182-1

Millar, R. J., Hepburn, C., Beddington, J., & Allen, M. R. (2018). Principles to guide investment towards a stable climate. *Nature Climate Change, 8*, 2–4. https://doi.org/10.1038/s41558-017-0042-4

Okereke, C., & Coventry, P. (2016). Climate justice and the international regime: Before, during, and after Paris. *WIREs Climate Change, 7*(6), 834–851. https://doi.org/10.1002/wcc.419

OLADE [Organización Latinoamericana de Energía]. (2021). *Panorama energético de América Latina y el Caribe 2020.*

Paris Agreement. (2015). (Adopted on 12 December 2015, entered into force 4 November 2016), United Nations Treaty Series vol. 3156 C.N.63.2016.

Pearse, R. (2017). Gender and climate change. *Wiley Interdisciplinary Reviews: Climate Change, 8*(2), e451. https://doi.org/10.1002/wcc.451

Quintero, R., & Silva, E. (1991). *Ecuador, una nación en ciernes, tomo 2: Economía.* Abya Yala. https://biblio.flacsoandes.edu.ec/libros/digital/58617.pdf

Radcliffe, S. (2012). Development for a postneoliberal era? Sumak Kawsay, living well and the limits to decolonisation in Ecuador. *Geoforum, 43*(2), 240–249. https://doi.org/10.1016/j.geoforum.2011.09.003

Rempel, A., & Gupta, J. (2020). Conflicting commitments? Examining pension funds, fossil fuel assets and climate policy in the Organisation for Economic Co-operation and Development (OECD). *Energy Research & Social Science, 69*, 101736. https://doi.org/10.1016/j.erss.2020.101736

Rempel, A., & Gupta, J. (2021). Fossil fuels, stranded assets and COVID-19: Imagining an inclusive & transformative recovery. *World Development, 146*, 105608. https://doi.org/10.1016/j.worlddev.2021.105608

Res4Africa. (2021). *Commercial and industrial renewable energy market for South Africa.* https://static1.squarespace.com/static/609a53264723031eccc12e99/t/6101131fdcb2d667b8cf0839/1627460412555/C%26I+Study_compressed+%282%29.pdf

Roelfsema, M., et al. (2020). Taking stock of national climate policies to evaluate implementation of the Paris Agreement. *Nature Communications*, 11. https://doi.org/10.1038/s41467-020-15414-6

Ross, M. J. (2012). *The oil curse: How petroleum wealth shapes the development of nations.* Princeton University Press.

Sawyer, S. (2004). Crude chronicles: Indigenous politics, multinational oil, and neoliberalism in Ecuador. In G. M. Joseph & P. M. Von Eschen (Eds.), *American encounters/global interactions*. Duke University Press. https://doi.org/10.1215/9780822385752

Scheyvens, R., Banks, G., & Hughes, E. (2016). The private sector and the SDGs: The need to move beyond "business as usual". *Sustainable Development, 24*(6), 371–382. https://doi.org/10.1002/sd.1623

SEI, IISD, ODI, Climate Analytics, CICERO, & UNEP. (2019). *The production gap: The discrepancy between countries' planned fossil fuel production and global production levels consistent with limiting warming to 1.5 °C or 2° C.* http://productiongap.org/

Sinn, H. W. (2012). *The green paradox: A supply-side approach to global warming.* MIT Press.

Stadelmann, M., Michaelowa, A., & Roberts, J. T. (2013). Difficulties in accounting for private finance in international climate policy. *Climate Policy, 13*(6), 718–737. https://doi.org/10.1080/14693062.2013.791146

Svampa, M. (2019). *Neo-extractivism in Latin America: Socio-environmental conflicts, the territorial turn, and new political narratives.* Cambridge University Press. https://doi.org/10.1017/9781108752589

UNCED [United Nations Conference on Environment and Development]. (1992). Rio Declaration on Environment and Development. A/CONF.151/26 (Vol. I). https://

www.un.org/en/development/desa/population/migration/generalassembly/docs/globalcompact/A_CONF.151_26_Vol.I_Declaration.pdf

UNEP [United Nations Environmental Programme]. (2019). *Emissions Gap Report 2019*. https://wedocs.unep.org/bitstream/handle/20.500.11822/30797/EGR2019.pdf?sequence=1&isAllowed=y

UNFCCC [United Nations Framework Convention on Climate Change]. (1992). (Adopted in 1992, entered into force 21 March 1994), 1771 UNTS 197 (UNFCCC).

UNFCCC [United Nations Framework Convention on Climate Change]. (2021, November 13). Decision -/CP.26 Glasgow Climate Pact. Advance unedited version. https://unfccc.int/sites/default/files/resource/cop26_auv_2f_cover_decision.pdf

UNGA [United Nations General Assembly]. (2015, September 25). *Transforming our world: The 2030 Agenda for Sustainable Development*. A/RES/70/1.

Urgewald. (2020, November 12). NGOs release the 2020 Global Coal Exit List: 935 companies that banks, investors and insurers need to avoid. https://coalexit.org/sites/default/files/download_public/GCEL2020Release_1.pdf

Valdivia, G., & Lyall, A. (2018). The oil complex in Latin America: Politics, frontiers, and habits of oil rule. In J. Cupples, M. Palomino-Schalscha & M. Prieto (Eds.), *The Routledge handbook of Latin American development* (pp. 458–468). Routledge.

Vásquez, P. (2014). *Oil sparks in the Amazon: Local conflicts, Indigenous populations, and natural resources*. University of Georgia Press.

Watts, M. J. (1999). Petro-violence: Some thoughts on community, extraction, and political ecology. Working Papers WP 99-1. Berkeley Workshop on Environmental Politics.

Welsby, D., Price, J., Pye, S., & Ekins, P. (2021). Unextractable fossil fuels in a 1.5 °C world. *Nature, 597*, 230–234. https://doi.org/10.1038/s41586-021-03821-8

World Bank. (2019, March 13). World Bank Group announcements at One Planet Summit. https://www.worldbank.org/en/news/press-release/2019/03/13/world-bank-group-announcements-at-one-planet-summit

WRI [World Resource Institute]. (2019). Stepping up NDCs: Resources for NDC enhancement, tracking progress. https://www.wri.org/ndcs

2. Analytical Framework: Inclusive Development, Justice and Energy Transition

Joyeeta Gupta, Barbara Hogenboom and Arthur Rempel

Abstract

Understanding how fossil fuels can be rapidly phased out in a just manner requires an analytical framework. This chapter combines an energy transition framework with an inclusive development framework. The energy transition is about how niche ideas can become regimes and then cause landscape transformation. Inclusive development focuses on social inclusiveness, ecological inclusiveness and relational inclusiveness. This chapter aims to create an analytical framework for this book (and future studies) to analyse whether ideas of energy transformations are in a niche or transformation stage and whether they are socially, ecologically and relationally inclusive. It also looks at the role of key actors—local groups (through using theories of social movements), investors and producers, and states in the phasing-out of fossil fuel and niche ideas that they are promoting.

Keywords: fossil fuel phase-out, socio-environmental justice, actors, inclusive development, transformation

2.1. Introduction

In order to analyse the role of different actors, arguments and approaches in leaving fossil fuels underground (LFFU), this book uses an analytical framework that combines several theoretical approaches. As explained in Chapter 1, we focus on the North–South dimensions of the urgent challenge

---

Gupta, J., Hogenboom, B., Rempel, A. & Olofsson, M. (eds), *Leaving Fossil Fuels Underground: Actors, Arguments and Approaches in the Global South and Global North*.
Taylor & Francis Group, 2024
doi: 10.5117/9789048560370_CH02

of phasing out fossil fuels and on the related global justice agendas as well as the policies and social movements in South Africa and Ecuador. Ranging from climate justice and socio-environmental justice to economic and intergenerational justice, these agendas emerge both from local social and political struggles, and from the global 2030 Agenda for Sustainable Development (Agenda 2030) (UNGA, 2015). Given the radical nature of this challenge and these agendas, it comes as no surprise that the social and political debates on LFFU involve a wide range of interesting ideas as well as profound differences over what the main obstacles and the best course of action and governance are. The academic studies and debates on the subject show similar diversity and dissonances. Our analytical framework acknowledges these differences and indicates the research direction through which we aim to identify successful strategies for LFFU and the possibilities of scaling them up. Section 2.2 discusses how our study of LFFU and energy transition relates to debates and concepts on inclusive development and socio-environmental justice. In Section 2.3, we examine a few relevant theories on the role of actors, approaches and arguments, focusing on social movements, investment actors and states. Finally, before drawing some conclusions, Section 2.4 integrates some key conceptual and theoretical elements into an analytical framework that will guide this book's overall assessment of the possibilities and obstacles for leaving fossil fuels underground globally, but especially in the Global South.

2.2. Energy transition and justice through inclusive development

2.2.1. Energy, development and socio-environmental justice

As any society and any type of development requires energy, the availability of energy has always been strategically important. Since fossil fuels became the main source of energy, countries with substantial reserves of coal, oil and gas were generally considered to be in an advantageous position for development. Provided that they could access the necessary technology and investment, they could use these energy resources for internal use as well as for exporting to countries that are less endowed. However, during the 20th century, it became clear that for societies in the Global South, having a wealth of fossil fuels does not necessarily generate substantial economic progress, social well-being and political stability. Beyond a variety of positive and negative examples, comparative analyses demonstrated that fossil fuel wealth can in fact even slow down countries' economic modernisation, poverty alleviation and democratic institution-building. While some call it the paradox of plenty (cf. Karl, 1997), this tendency has become best known as the "resource curse" (cf. Ross, 1999). While fossil fuel extraction and exports generate massive foreign investments and state revenues, in Africa, Latin America and beyond, this often perpetuates uneven development instead of contributing to inclusive development (Bebbington et al., 2018a; Hogenboom, 2012; Hogenboom, 2018; Rempel & Gupta, 2021).

Academic studies into the nature and causes of the resource curse in the Global South point to several problems and perverse effects related to fossil fuel wealth. Dependency on external technology, markets and knowledge implies dependency on foreign companies, banks and governments. While most countries nationalised their fossil fuel reserves and created state-owned companies to control part of the operations (see 2.3.4), this external dependency remained prominent. The fact that fossil fuels are connected to large capital flows may generate even larger risks. Economically, the dependency on revenues with volatile prices is problematic, as is the dependency on powerful foreign companies. Furthermore, large-scale fossil fuel exports usually harm other sectors and economic diversification. While this tendency is commonly known as the "Dutch disease," Coronil (2008) prefers to call it the "neocolonial disease." Financially, the availability of large fossil fuel reserves triggers the accumulation of foreign debts, not only to allow for the development of fossil fuel projects and infrastructure, but also because these reserves are viewed as a collateral for additional lending by both the national government and external creditors. Politically, fossil fuel revenues are prone to give way to inefficient public spending, corruption and elite capture. It also tends to trigger a culture of rent-seeking behaviour that affects the public sector, the private sector and organised civil society (e.g. trade unions), resulting in the development of a rentier state and even a rentier society (Peters, 2017). Moreover, geopolitically, large national reserves and exports are automatically of huge international importance, involving actors and interests that easily overpower national actors and interests. Even in countries of the Global South where governments recently claimed to break away from the past and use fossil fuel wealth for the benefit of inclusive development and national sovereignty, such as in Latin America, many of the old patterns and problems were repeated (Svampa, 2015).

Particularly perverse "side effects" of fossil fuel wealth in the Global South can be found at the local sites of extraction, where communities and nature experience the disruptive and destructive side of this industry. While consciousness about the global problem of climate change and the need to quickly make an energy transition has increased, consciousness about the local impact of the extraction, upgrading and transport of fossil energy sources at numerous sites remains relatively limited. The social and environmental externalities tend to be particularly huge in remote and marginalised areas, where not only the industry's technology, financial and human capital are more limited, but also governmental oversight is weak, and activists may have to risk their lives to rightfully protect their health, income, land and livelihoods. Human rights, ecological and distributional dimensions of the fossil fuel sector continue as a result of both the lack of visibility—even to a large part of society in the countries themselves—and the large national and international interests at stake (O'Rourke & Connolly, 2003; Watts, 2005). In short, environmental and social justice are often simultaneously impaired by the fossil fuel industry (Sovacool, 2016).

Despite the inequalities that the fossil fuel enterprise creates, it also fosters multiple dependencies: from the poorest to the richest in society, they all need fuel for their livelihoods and jobs. Ironically, the poor often depend more on cheap fossil fuels. This creates an additional challenge for achieving an inclusive energy transition. Moreover, the need to address energy poverty also ends up being used as a justification by states and energy companies to continue to invest in fossil fuels. Simultaneously, the huge amount of capital involved in the fossil fuel sector make it a lucrative business for the Global North to not only use the Global South as a source of resources (à la dependency theory) but also to dump its know-how, technology and infrastructure onto the Global South now that it is clear that fossil fuels need to be phased out (see Chapter 9).

Socio-environmental justice can be seen from two different schools of thought. The legal school has engaged with justice issues for centuries, focusing on how justice can be defined and how it evolves over time. Where legislative systems fail to address justice issues, the court systems are critical for testing justice concepts within a specific context. Legal systems do not (necessarily) promote a certain kind of justice; they evaluate different claims to justice and weigh these against each other. More recently, the political ecology school of justice focuses on how social dimensions (e.g. in relation to race, ethnicity, gender, class, caste, deprivation, disability, age, generation) influence and are affected by development strategies and other political decisions, and their environmental aspects—air, water and land pollution, loss of biodiversity, loss of fisheries and so on. Political ecology focuses on how marginalised and vulnerable groups are exposed to major intersectional inequalities through the way the society and economy are designed (Walker, 2012). Political ecology studies focus on economic, ecological as well as cultural distribution conflicts (Escobar, 2006). They demonstrate "a sensitivity to environmental politics as a process of cultural mobilisation, and the ways in which such cultural practices—whether science, or traditional knowledge, or discourses, or risk, or property rights—are contested, fought over, and negotiated" (Watts, 2000, p. 259). Social mobilisation over environmental concerns not only serves immediate local needs but simultaneously challenges systemic inequalities. The two schools can be seen to intersect on the streets and in the courts. When social movements challenge existing policies in courts, the courts weigh these arguments against those of the other actors. As courts may be conservative in trying to promote a predictable and fair legal system, initial court cases by social movements may fail. But as social movements persist, the values and norms they promote acquire greater legitimacy and courts may accept their position and this leads to a new equilibrium in justice values in a society. In this book, we look at how both social movements and courts are addressing and resolving justice issues around energy.

For the study of LFFU we also need to look into debates and theories on climate justice and climate debt. Following directly from the environmental justice movement, climate justice "addresses the intersections of 'green' ecological concerns and 'red' socio-economic ethical considerations, ranging from public policy deliberations to political practices" (Bond, 2013, p. 133). Climate justice advocates tend to be critical of UN interstate climate negotiations for insufficiently dealing with the current and historical responsibility of the largest GHG-emitting countries, the major transnational energy companies and the Global North. Although in 2012 the concept of climate debt was ultimately recognised by the UNFCCC, putting climate debt compensation or payments into practice is still a challenge. Since the early 1990s, there have been proposals to address liability and climate debt. More recently, interesting, detailed proposals have been developed in Latin America (Ecuador) and Africa (Namibia) (Bond, 2018). In addition, when it comes to climate change and socio-environmental justice, a particularly challenging issue to tackle concerns state–corporate crime: "the ways illegal acts and social injuries often emerge from the intersection of economic and political power" (Kramer, 2013, p. 157). This short overview shows that political economy is also central to academic studies and debates on energy, climate change and socio-environmental justice and requires transformative change of the state and corporations. It also raises issues regarding what kind of development we want.

As the following subsection (2.2.2) will indicate, our multidimensional approach to justice resonates with the concept of inclusive development.

Social and ecological inclusiveness refers to the challenges faced by those externalised or marginalised by existing policies, while relational inclusiveness, inter alia, looks at the relations between actors and how social movements fight for justice and how courts and legislatures respond. Our multi-level approach to justice will focus on the differing justice issues at all levels of governance.

2.2.2. Inclusive development: A comprehensive multidimensional perspective

The question of what kind of development should be strived for has been heavily discussed for centuries and links to fundamental social, political and academic debates. Since the mid-20th century, the debates about how to overcome development problems in the Global South have generated numerous new views and policy agendas. Many of these departed from an economic and social development objective—for example, dependency and underdevelopment theories that proposed fundamentally restructuring the global political economy and argued for national and global agendas for poverty reduction and pro-poor growth that aimed at policy reforms and additional funding. Subsequently, ecological crises and the awareness of global environmental limits and interdependencies gave way to new views of how to develop within ecological boundaries, both in the Global North and the Global South. Ideas on sustainable development initially stressed the importance of including economic, social and environmental goals; current and future generations (WCED, 1987). While sustainable development policies of multilateral institutions have been criticised for only addressing socio-environmental needs to the extent that it would not harm dominant economic and political interests (and consolidating rather than overcoming global inequalities and injustice), the idea or at least the discourse of sustainable development gained wide support from NGOs, businesses and governments and led to the adoption of the Sustainable Development Goals (SDGs). Increasingly, calls for overcoming gender inequality and racism also gained strength and have been included in global debates and agendas for sustainable and inclusive development, such as the Millennium Goals and the SDGs. Since the turn of the century, however, the ways in which new concerns and adjectives were adopted in development discourses and policies of major international actors have been criticised for not addressing the root causes of the problems and for, at best, only reforming rather than transforming the global system. In particular, the growing awareness of climate change and the loss of forests and biodiversity have sparked new debates on degrowth and alternatives for development (such as *ubuntu*, *buen vivir*, gross national happiness [Van Norren, 2017]), calling for a fundamental rethinking or even altogether leaving aside the idea of development.

In this book, we adopt an inclusive development perspective where inclusiveness is *not* an adjective to development—instead, it redefines the content of development. Core to this approach is the notion of socio-environmental and relational justice. This multidimensional and comprehensive approach thus enables us to look at the problem of phasing out fossil fuel with its huge economic costs from a social, ecological and relational perspective (Gupta et al., 2015; Rauniyar & Kanbur, 2009; Rammelt & Gupta, 2021). This is different from adopting a sustainable development perspective as even though strong sustainability focuses on minimising economic, social and ecological trade-offs, the economic values still dominate as other dimensions cannot (and probably should not) always be translated into monetary terms.

Before discussing this comprehensive approach to inclusive development, let us indicate why it is important by referring to the debate around stranded resources/stranded assets. The stranded resources/assets theory points out that a phase-out of fossil fuels will imply that fossil fuel infrastructure and assets can no longer be used and that a substantial share of the fossil fuel resources and reserves become stranded resources (Griffin et al., 2015; Linnenluecke et al., 2015; Rozenberg et al., 2015). Based on the calculation that to achieve the 2 °C target, only 565 of the 2,975 gigatonnes (so less than one-fifth) of CO_2 available in assets can be used (McKibben, 2012), this results in huge, stranded assets and resources. Asset and resource holders will be reluctant to give these up (Matikainen & Soubeyran, 2022; Ansar et al., 2013), but they are in essence facing a financial risk—referred to as the carbon bubble (Carbon Tracker Initiative, n.d.). Whether phasing out fossil fuels happens due to new policies and legislation (e.g. carbon pricing), pressure from social movements and/or a market reaction (e.g. a shift towards renewables), it will require the fossil fuel industry and those who invest in it to reduce their reliance on fossil fuels (Ansar et al., 2013). As the economic costs and financial risks of leaving fossil fuels underground are very high (see Chapter 1), actors that mainly take economic arguments into account will be unable and unwilling to prioritise the ecological and social aspects. In effect, the economic approach, arguments and actions work against a rapid and fair phasing-out of fossil fuels.

An inclusive approach, in contrast, prioritises the access to a minimum of resources for all people, and a just allocation of the remaining resources, risks and responsibilities, both for current and future generations. Hence, a comprehensive inclusive development approach allows us to take a social, ecological and relational perspective on economic issues, including challenging issues such as stranded assets. Chapter 1 presented some of the energy-related justice elements that also emerge from Agenda 2030. We identified socio-environmental justice issues such as access to income and services, but also to procedural rights and nature's contributions to people. Equally important is a fair allocation of energy and other resources through reducing inequality and changed production and consumption patterns within and between countries. Evidently, energy and its infrastructure are also indispensable for the economy.

The comprehensive inclusiveness approach applied in this book has three dimensions: social, ecological and relational (see Table 2.1). *Social inclusiveness* (justice) focuses on access issues. It includes access to (i) income/jobs above the poverty line; (ii) access to water, food, energy, health, housing and infrastructure; and (iii) access to procedural rights, including non-discrimination, rights of minorities (including Indigenous and spiritual communities) and women; rights to information, decision-making, recognition and courts. In terms of allocation, social inclusiveness is about sharing resources, risks and responsibilities equitably (Gupta & Lebel, 2020; UNDP, 1994). As mentioned in Chapter 1, Agenda 2030 includes all these access issues related to social inclusiveness. It also stresses that inequality must be addressed at all levels of governance, including globally, including income, production and consumption patterns and the impacts on least developed countries and small island states.

Ecological inclusiveness in terms of access is about access to the beneficial aspects of nature's contributions to people (NCP) (Díaz et al., 2018) and reduced exposure to ecological risks for the most marginalised and vulnerable people. Such access enables income generation from nature and reduced health risks from pollution, and it is protected through environmental rights and nature conservation as well as through territorial rights. In terms of allocation, ecological inclusiveness is about how natural resources, risks and responsibilities are shared, and how natural systems are protected, from the local to the global level. Here, citizens' rights in environmental decision-making are key ingredients contributing to environmental policies and agreements. Principle 10 of the Rio Declaration on Environment and Development (UNCED, 1992) is a global commitment to grant every citizen the right of access to information, to participation in decision-making and to justice in environmental matters, which has been materialising at the local to regional level. So far, the Pan-European Aarhus Convention on Access to Information, Public Participation in Decision-making and Access to Justice inEnvironmental Matters (Aarhus Convention) (1998) is the most ambitious step towards environmental democracy undertaken under the auspices of the UN (Ebbesson, 2007, p. 686), but in 2021 the Regional Agreement on Access to Information, Public Participation and Justice in Environmental Matters in Latin America and the Caribbean (Escazú Agreement, 2018) entered into force, which is expected to improve the protection of environmental defenders and vulnerable populations (Stec & Jendrośka, 2019).

Table 2.1. Inclusive development with socio-environmental justice

Inclusiveness	Just access to:	Just allocation of:
Social	(i) income/jobs above the poverty line; (ii) access to water, food, energy, health, housing and infrastructure; (iii) access to procedural rights, including non-discrimination, rights of minorities, women, developing countries; rights to information, decision-making, recognition and courts	(i) socio-economic resources (e.g. fossil fuel resources) and infrastructures; (ii) socio-economic risks; (iii) socio-economic responsibilities
Ecological	NCPs—material, non-material, regulating (all with a cultural component)—for the most marginalised to enable (i) income; (ii) healthy lives; (iii) environmental rights	(i) remaining NCPs; (ii) related risks; (iii) environmental responsibilities
Relational	Role of states (and interstate relations) and their ability to provide and/or protect social and ecological common goods and to address inequalities; role of other actors, including financial actors and their desire for maintaining the status quo, and social movements and their desire for change; role of dominant and alternative approaches and arguments in shaping the adoption of substantive and procedural principles, rights and responsibilities to address socio-environmental justice.	
Development	Inclusiveness redefines development away from growth to prioritise socio-environmental well-being of current and future generations	

Relational inclusiveness focuses on the politics of inclusiveness with two ontological strains; it examines existing inclusiveness/exclusiveness from the perspective of political ecology and subsequently explores means to address political exclusiveness from a social constructivist angle. Relational inclusiveness focuses on the role of the state, interstate relations and state–industry–people relations in providing and/or protecting public and merit goods; and the way discourses are translated into principles and instruments for socio-environmental justice. It looks at how other actors shape this relationship and in particular both social movements as well as large financial actors. Relational inclusiveness recognises that resource shortage within a neoliberal capitalist world leads to decisions in favour of commodifying and privatising resources and promoting private sector participation in the supply of merit and public goods as well as using modern technologies to avoid or minimise the application of tax and labour laws.

Relational inclusiveness recognises that in an international relations context, control over resources—and hence control over the factors of production—leads to hegemonic approaches where states use sovereignty (see the adoption of the "full permanent sovereignty" argument in Chapter 1) and securitisation to justify not sharing resources, not accepting responsibility for damage caused to others and proactively using neoliberal capitalist strategies to access resources elsewhere (e.g. land and water grabbing) (Zoomers, 2010; Fairhead et al., 2012). These dominant tendencies stand in the way of a transition to inclusive development. Relational inclusiveness thus questions the perpetuation of non-inclusive systems and calls for more accountable governance systems that pre-empt the accumulation of power and wealth through dispossession which benefit some while harm others as well as nature (Gupta et al., 2015; Harriss-White, 2006; Hickey et al., 2014; Mosse, 2010).

Ultimately, inclusiveness redefines the content of development by arguing that development can no longer be measured in terms of GDP growth and requires a transition towards inclusive wealth (UNEP, 2018) or other ways of calculating human and ecosystem well-being. For the purpose of this book, our inclusive development perspective implies that our units of analysis include the issues presented in Table 2.1.

2.2.3. An inclusive development perspective on energy transition

Inclusive development is the normative lens that we apply to the challenge of phasing fossil fuel out. This challenge is often discussed in terms of the energy transition which refers to a supply-side process of moving from high-GHG-emitting energy sources to low- to no-GHG-emitting energy sources. This means moving from coal to oil to gas to renewables (with or without nuclear energy and biofuels). Such a transition can be incremental moving upwards from one level to the next or transformational where one skips certain levels. However, next to the entire supply side of infrastructure from extraction, production and distribution, for a full transition also the demand side of infrastructure and technology from households to industrial users will have to change (Geels et al., 2016).

In understanding the energy transition, it is important to appreciate what a transition is. Transitions are "processes of structural change in major societal subsystems" (Meadowcroft, 2009, p. 324) that "profoundly [alter] the way [society] functions" (De Haan & Rotmans, 2011, p. 92). A transition often requires that all kinds of societal systems have to simultaneously change—this includes cultures and social systems, but also infrastructures as well as investments; and these have to change at all levels of governance, especially when the underlying energy system is a globally embedded system (Geels et al., 2016). This requires alignments not only between society, politics, economics and infrastructure, but also between "technologies, supply chains, infrastructures, markets, regulations, user practices and cultural meanings" (Geels, 2018, p. 224).

A transition normally requires experimental *niches* which can lead to changes in *regimes* before the entire *landscape* can change. This requires leaders and policy entrepreneurs, investors and an encouraging policy and legal infrastructure. It also requires the identification of micro-level technological niches that are radically different from the existing system and which can be nourished as incubation rooms and which can be scaled up to larger geographical areas and can lead to mainstreaming at the end (Geels & Schot, 2007). In the process several stages are distinguished: *pre-development* (status quo, small changes, niches; often involves high risk; can result from the work of NGOs or protests by social movements); *take-off* (collaboration between actors facilitated by windows of opportunity); *acceleration* (changes in institutional dynamics); and *stabilisation* (a new system is established) (Rotmans et al., 2001; Kemp & Loorbach, 2003). In order to stabilise, there needs to be a new status quo achieved where different government, businesses and civil society actors within the "triangle of governance" (Abbott & Snidal, 2009) come together and develop "shared cognitive routines" supported by policy (Geels & Schot, 2007). Figure 2.1 portrays what the LFFU transition would look like, according to these transition theories following the pre-development phase.

However, transition processes face major bottlenecks. Sociocultural bottlenecks refer to the reluctance of society to change or in our story the way in which power politics determines what happens. Technological and infrastructural bottlenecks generally refer to lock-ins within which society functions—where the infrastructure and related technologies promote a certain kind of behaviour and it is difficult to change social behaviour without changing the technological and infrastructural system in society. There is also intense debate about the extent to which the appropriate technologies for a full energy transition are available and can be used. All this is embedded in a political context where values such as an emphasis on economic growth and other related values create a broader ideological lock-in (Geels, 2002).

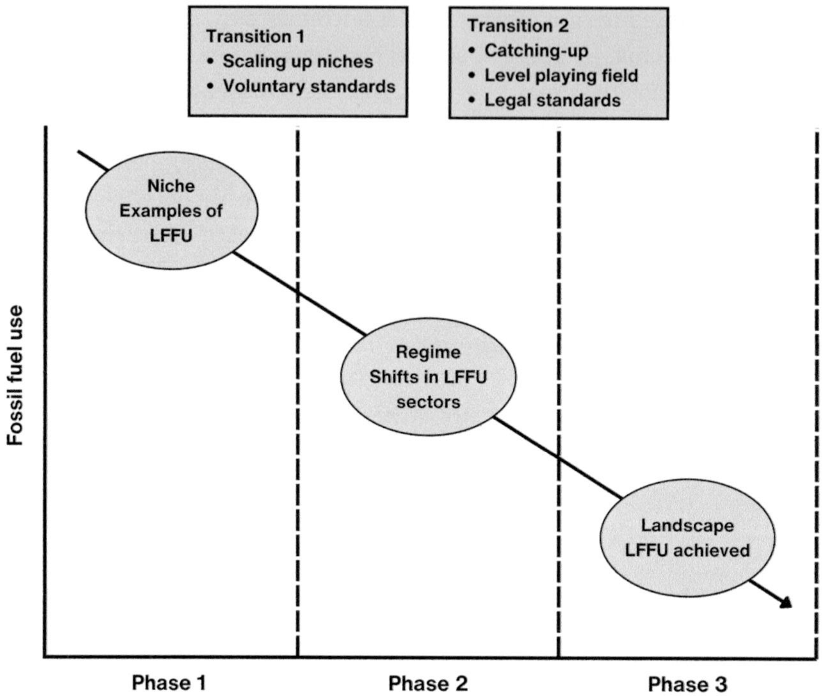

Figure 2.1. Transition theory applied to LFFU following the pre-development phase.

To any innovative institutional and societal change, including those related to natural resource extraction and use, the politics of ideas are of key importance (Bebbington, 2013). Ideas about how to overcome major problems and injustices are abundant but understanding which new ideas and arguments can be persuasive and can overcome vested interests is the first step to actually develop, realise and up-scale alternatives. Ideas that affect policymaking can take different shapes, ranging from programmes and paradigms at the cognitive level, to frames (including symbols and concepts) and public sentiments at the normative level (Campbell, 1998). While experts and epistemic communities influence public discussions and policies and thus serve as agents of resource governance change, more often "ideas become influential when they are bundled with movements and coalitions" (Bebbington, 2013, p. 10). Thus, it is the *politics* of ideas that play an important role in changing dominant patterns, and this involves multiple actors and scales. When we link this to what the LFFU transition could look like (Figure 2.1), we can appreciate that the politics of ideas are relevant throughout the process: from the implementation of stepping-stone experiments to scaling up and the shifts in regime and even landscape. In

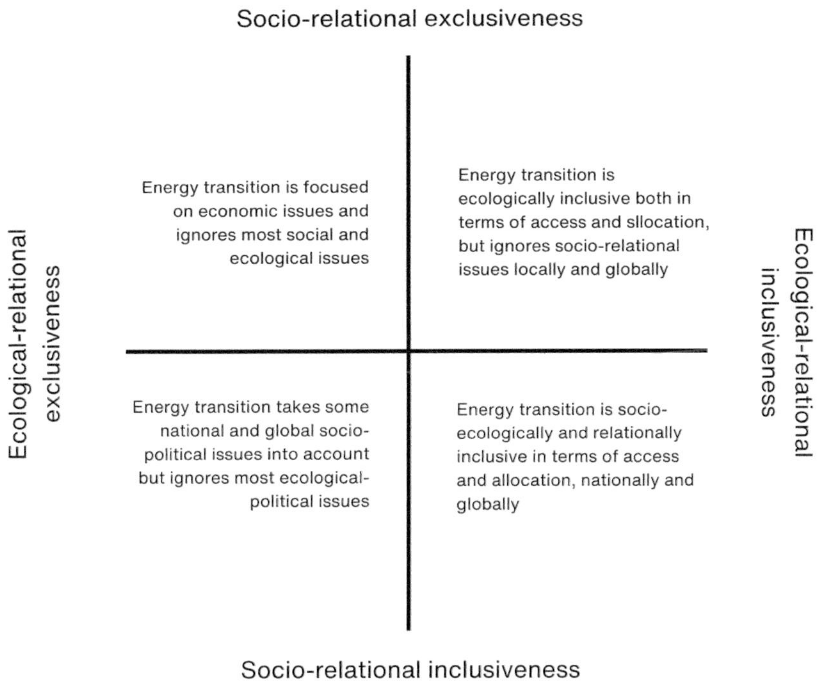

Figure 2.2. Inclusive development and the energy transition.

addition, for new ideas to become persuasive and implemented, LFFU actors, arguments and approaches also need to address the (potential) bottlenecks of transition, especially sociocultural and ideological bottlenecks. Considering the magnitude of the LFFU transition, it is evident that only approaches and arguments that include a wide range of actors, interests and needs are likely to realise all transition steps and effectively leave fossil fuels underground. Therefore, we argue that combining the LFFU transition with inclusive development and national changes with global changes reflects both a normative agenda and political pragmatism.

When we combine our ideas on the energy transition with our approach to inclusive development with socio-environmental justice, we get a figure consisting of four quadrants (Figure 2.2). Only in one quadrant is the energy transition to LFFU socially and ecologically just at the local to global level. In this book, we examine how these transitions are materialising and which actors, arguments and approaches can make them more socio-ecologically and relationally inclusive. In the following chapters, we explore the two national cases of South Africa and Ecuador, but also transition elements within the Global North and South at large. The chapters identify niches and key actors (social movements, grassroots organisations, financial institutions and state institutions) and their arguments and approaches, and explores how these niches can be scaled up to prompt regime (and eventually landscape) change.

2.3. Theories on actors, approaches and arguments

2.3.1. Introduction

From the perspective of relational inclusiveness, a key element for realising a fair energy transition—that is, an energy transition with inclusive development and socio-environmental justice—is understanding the role of actors, arguments and approaches. This section reviews three key actors: Social movements and their role in pushing for socio-environmental justice; large companies and investors and theories that discuss their role in maintaining the status quo; and the role of the state and interstate relations in addressing challenges, such as leaving fossil fuels underground.

2.3.2. Social movements

Social movements refer to a coming together of large groups of people and groups who coalesce around common grievances and claims for justice and rights. They can be mobilised by a few core (non-governmental) organisations and/or inspiring persons (e.g. Greta Thunberg). Social movements are flexible and loosely coordinated, and their ability to sustain over a large period of time depends on the degree of grievance and their capacity to generate narratives (e.g. claims, arguments and proposals) that mobilise a substantial number of citizens and capture the public imagination. As their ideas, actions and networks often transcend national and continental boundaries, large movements are transnational by nature. An early wave of social movements focused on opposing the monarchy (e.g. the French and Russian revolutions) and colonisation, leading to the creation of republics and decolonised states. Much of these movements were also around class struggles (Calhoun, 1993; Melucci, 1996). The labour movements demanded workers' rights, which led to the institutionalisation of such rights in laws and policy, whereas peasant movements sometimes succeeded in their demands for land distribution. As of the 1970s, social movements emerged centring more on identity-related issues and rights, such as human rights, feminism and women's rights, non-discrimination, rights of Indigenous peoples, sexual and reproductive rights, and the rights of LGBTQ (lesbian, gay, bisexual, transgender and queer) communities. In parallel, the environmental movement has developed momentum. Together, the worldwide rise of these movements has contributed to the recognition and protection of various rights.

Since social movements result from an autonomous mobilisation of people around a specific theme, the question arises: What leads to the origin of a movement and when and under what circumstances can it be successful in promoting change? Resource mobilisation theory (McCarthy & Zald, 1977; Tilly, 1978) focuses on the role of intellectual, financial and organisational resources in mobilising people to coalesce around a strategic narrative (Diani & Bison, 2004). New social movement theory (Larana et al., 2009; Zald, 2000) focuses on the importance of generating a shared ideal, identity or narrative that binds people together to challenge dominant ideas, attitudes and actors (McAdam et al., 2001; Snow et al., 2019). In addition, grounded theory has been influential, building on evidence collected from different case studies of social mobilisation (Strauss et al., 1998, p. 12; Edelman, 2001, p. 285).

Environmental movements have developed all over the world, but they originated through different agendas in different places. Nature conservation represents one early expression, especially but not exclusively in the Global North, while peasant and Indigenous struggles to protect land, water and forests, have been a more common root in the Global South. More has been registered in the academic literature of the rise of the environmental movement in the developed world, but there have been strong movements in India and Latin America since the 1970s. The movements in India focused on the impacts of large dams (such as the social movement Narmada Bachao Andolan [Save the Narmada River]) and were against deforestation (the Chipko movement). The environment justice movement in its current form emerged in Latin America. These movements recognised that social and ecological issues were interlinked and had to be addressed together (Acselrad, 2010; Buttel, 1992) and that local struggles were intimately related to multi-level politics, including globalisation (Martinez-Alier et al., 2016; Temper et al., 2015, p. 256).

Many environmental movements are now focusing on issues related to fossil fuels. Extinction Rebellion and Fridays for Future are global movements on climate change, but there are many more and these are discussed in more detail in Chapter 3. And as we will see in Chapters 4–8, interesting new approaches and arguments regarding fossil fuels, development and nature have been developed in the Global South, such as anti-extractivism, living well and rights of nature. In this sense, one of the social movements' most powerful tools for both autonomy and influence arguably lies in "the art of organising hope": Indigenous and non-Indigenous movements have proven a great creative capacity to think beyond existing frameworks of the not-yet and prefigure alternative realities (Dinerstein, 2014). However, the social movements for climate justice also suffer from a division between grassroots climate justice activists, who portray such a radical thinking, and established international NGOs (Bond, 2018); there are differences between groups in the Global North and Global South; and different social movements focus on different themes (see Chapters 4 and 6).

Next to direct action, advocacy and lobbying, socio-environmental movements use litigation to achieve their goals. Their success in using litigation depends on the legal opportunity structure, which in turn depends on access to justice (which allows for legal standing, affordability), legal stock (precedents, justiciable rights and laws that may help) and judicial receptiveness (where judges are willing to engage in judicial creativity and activism) (Hilson, 2002). This structure focuses on the factors that influence whether and where social movements use courts to demand policy change (Vanhala, 2012; Wilson & Cordero, 2006). Use of courts can lead to an individual judgement which sets a precedent for other such cases, or it can lead to policy change at national level (Vanhala, 2018; De Fazio, 2012). The use of courts is also increasingly happening in the Global South (e.g. Setzer & Benjamin, 2020; Setzer & Vanhala, 2019).

2.3.3. Investors and producers

The ways in which investors and producers (companies) engage in energy governance often challenges the energy transition and relational inclusiveness. In the context of neoliberal capitalism, producers and investors have focused heavily on deregulation at national and international level and on self-regulation or self-governance and public private partnerships. Despite decades of attempts to develop a UN code of conduct for transnational corporations, this has not succeeded. Companies have instead been able to pre-empt regulatory processes by suggesting that they will undertake to regulate themselves. Fuchs (2013) argues that companies have instrumental, structural and discursive power. Instrumental power refers to their ability to lobby policymakers because they have easy access to them due to: their financial power and because government resources are increasingly dwindling; their ability to get a seat at the table in advisory capacities; and their ability to convene round tables to which they invite the government. Structural power refers to their huge financial resources, their ability to shift short-term production contracts to any part of the world to exploit areas with low governance capacity and their growing role in making rules and setting standards. Discursive power refers to how they use public campaigning to promote themselves and to promote images on the green economy and corporate strategies. Companies increasingly use power relationally to control the state and other actors (Fuchs, 2013).

For our study on LFFU, corporate self-regulatory trends are of great relevance, in particular, corporate social responsibility and socially responsible investment (SRI). With respect to the environmental dimensions, these trends are part of the approach of green economy or green business, which copies part of the ecological discourses but aims at preventing state regulation by letting corporate actors themselves be in charge. Corporate social responsibility is a concept that expects companies to take responsibility for their behaviour with respect to the people they employ and the people they affect. It requires them to take the social and ecological implications of their business into account. This includes ethical business practices, proper labour conditions and wages, and ecologically sound strategies with respect to the resources used and wasted. Corporate social responsibility code books list initiatives that many transnational fossil fuel companies adhere to, such as the Organisation for Economic Co-operation and Development (OECD) Guidelines for Multinational Enterprises, the Global Sullivan Principles of Social Responsibility, the UN Global Compact, the Voluntary Principles on Security and Human Rights, the Tripartite Declaration of Principles Concerning Multinational Enterprises and Social Policy, ISO 140001, the Global Reporting Initiative, and many more. Today there is a plethora of self-regulation standards in and for businesses, but these have been mostly used to pre-empt regulation and many argue that this is a form of greenwashing.

Increasingly investment companies are also presenting themselves as acting responsibly through impact investing and SRI. In recent years, the concept of impact investing has become important. It aims at ensuring that companies invest taking into account some specific social or ecological issue on which they wish to make an impact. Much of the literature is in the grey literature realm (Maduro et al., 2018). SRI examines the investment patterns of large investors where profit maximisation is made subject to the creation of "blended value" (Bugg-Levine & Emerson, 2011; Clarkin & Cangioni, 2016; Nicholls & Pharoah, 2008), which enables the creation of other kinds of values that emerge from addressing existing problems. The deregulation process enables business to avoid responsibilities caused by externalising social and ecological harm and they were then able to reinvent themselves as aiming for the broader welfare of society. This was often a response to e.g. the anti-apartheid movement in South Africa and the environmental movement more broadly (Bugg-Levine & Emerson, 2011). However, blending social responsibility with profit had to be financed in some way and this led to greater convergence with the aid sector (Bugg-Levine & Emerson, 2011; Clarkin & Cangioni, 2016; Sardy & Lewin, 2016; Nicholls, 2010; Hochstadter & Scheck, 2014; Nicholls & Pharoah, 2008). Impact investing is often linked to the Global Impact Investing Network, Impact Reporting and Investment Standards, and the Global Impact Investments Rating System. A key question that emerges in the literature is how effective these organisations are in doing what they claim to do and to what extent is this all "greenwashing" or "impact washing."

2.3.4. States

Last, but evidently not least, the state and interstate relationships play a key role in energy governance and the ways in which the use and fading out of fossil fuels is deeply entangled with development and justice. Related to relational inclusiveness, we focus on theories on the role of the state in the provision and protection of public and merit goods, and how neoliberal capitalism stands in the way of these tasks. In terms of interstate relations, we look at the role of sovereignty and the no harm principle, the right to (sustainable) development, and the role of states in international financial mechanisms. Public goods are goods that are non-exclusive (no one can be excluded) and non-rival (one's use does not mean there is less for others). The climate and hydrological systems are public goods. Generally, public goods need to be maintained by collectives—communities or states. Merit goods are goods that the public should consume but may be unwilling to do because of the costs involved—e.g. education, sanitation services, etc. (Kaul et al., 2004). This is also something that should not be privatised or commodified. However, states do not always work for the public and democratic societies have not always matured in different parts of the world. There are many debates on these issues—and we will return to them in the case study chapters.

Despite a strong neoliberal tendency among states to facilitate rather than regulate corporate actors, also resource nationalism continues to be a strong factor in the approaches and arguments on fossil fuel governance, especially among states from the Global South.

> Resource nationalism is the notion that natural resources should be managed above all for the needs of "the nation," and therefore should

> be controlled by the state for "the people." This idea is closely related to historical experiences of colonialism, dependent development, and the capture of resource rents by foreign capital. (Bebbington et al., 2018a, p. 207)

Following independence, many post-colonial societies felt that the global system was unfair and that they were still being held back and marginalised by trade and investment laws and policies. This led to a demand by the developing countries' states for permanent sovereignty over their natural resources in the 1970s, which has been reiterated in Agenda 2030.

Yet also in the Global North, the state and state-owned companies are key actors in the fossil fuel sector, as the governance of oil illustrates most clearly. Oil is generally much more embedded within state structures than other commodities. Despite transnational investments and powerful corporations, the global oil sector has an exceptionally high degree of state interference, and most (80% to 90%) reserves are under control of state-owned companies (Bridge, 2008). As countries still heavily depend on this energy source, the availability and price of oil determines economic performance, and control over oil is key in (military) security and geopolitical affairs. In global production networks states take on various roles in upstream (exploration and extraction) and downstream (refining and distribution) activities: the state is usually the owner of the resource, which grants concessions for extraction; many operating companies are at least partly state-owned; states are key regulators of the production chain (e.g. taxation, safety and environmental legislation); with foreign loans of development banks, states also engage in major investments in infrastructure to incentivise the oil sector; and they heavily influence the prices for consumers through policies, taxes and subsidies. State actors of both the Global South and the Global North have a key role in leaving fossil fuels underground as part of a global agenda on climate change.

2.4. The analytical framework in a nutshell

This book uses a comprehensive inclusive development lens which is operationalised to assess the challenge of LFFU and stranded assets as part of the energy transition. We are also informed by theories on the role of different actors (in particular, social movements, investors and states) in promoting LFFU, the kinds of motivations they have, the arguments they use and the approaches they develop to promote or resist LFFU. Next to their arguments and approaches, the motivation and relative powers of these

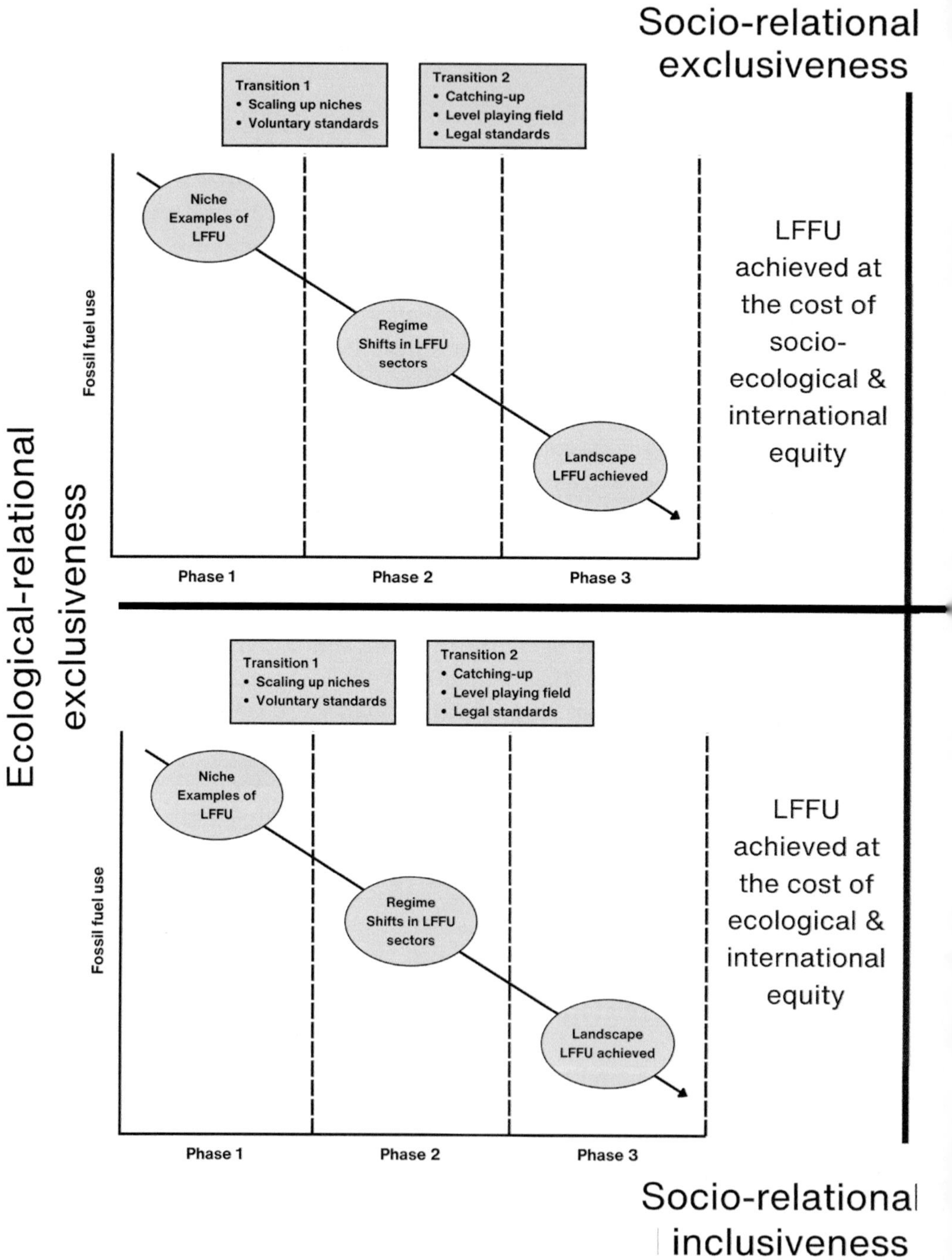

Figure 2.3. Model linking inclusive development and LFFU transitions.

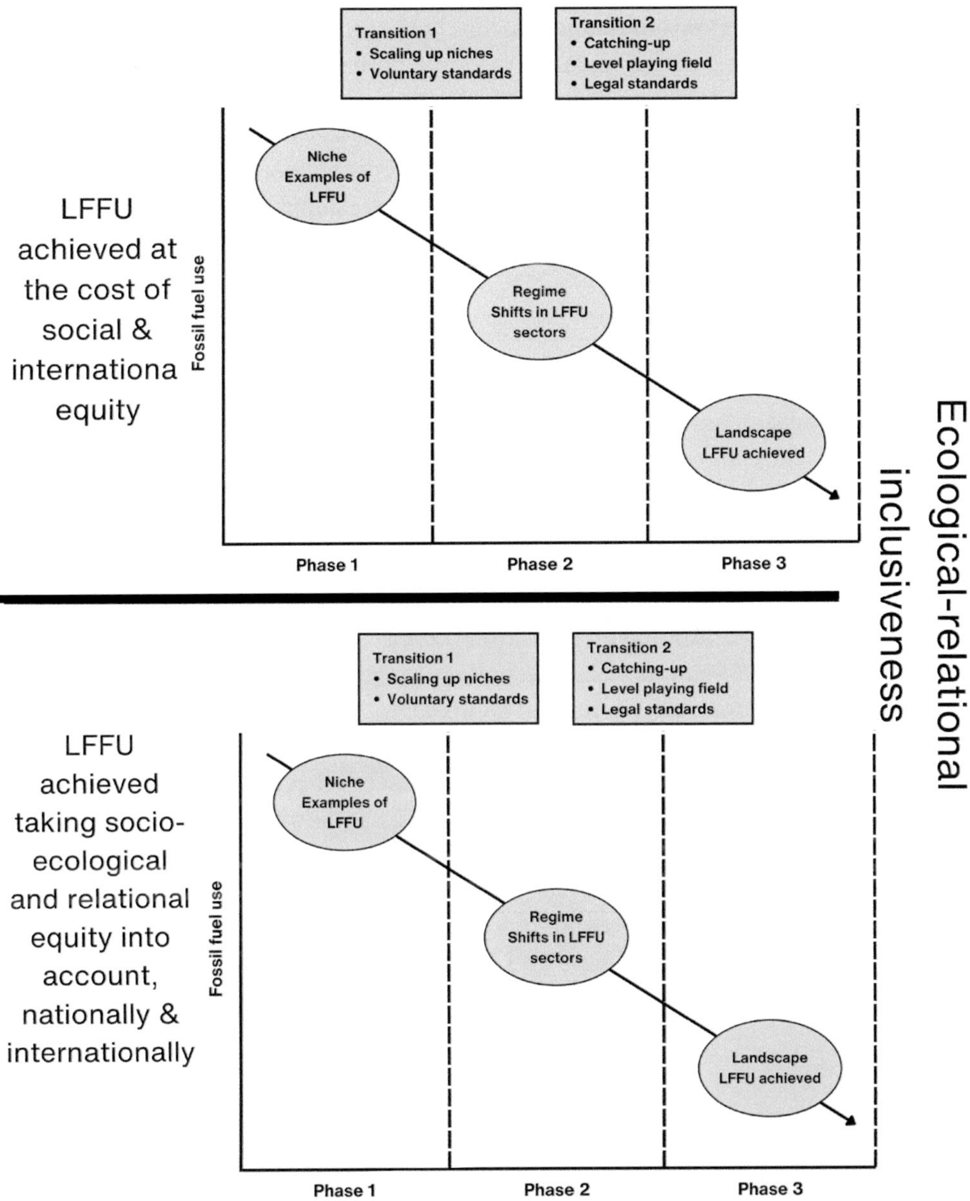
Transition 1
• Scaling up niches
• Voluntary standards
Transition 2
• Catching-up
• Level playing field
• Legal standards
LFFU achieved at the cost of social & internationa equity
Fossil fuel use
Niche Examples of LFFU
Regime Shifts in LFFU sectors
Landscape LFFU achieved
Phase 1
Phase 2
Phase 3
Transition 1
• Scaling up niches
• Voluntary standards
Transition 2
• Catching-up
• Level playing field
• Legal standards
LFFU achieved taking socio-ecological and relational equity into account, nationally & internationally
Fossil fuel use
Niche Examples of LFFU
Regime Shifts in LFFU sectors
Landscape LFFU achieved
Phase 1
Phase 2
Phase 3
Ecological-relational inclusiveness

actors strongly affect their influence. All these elements are also relevant for assessing if and how individual ideas (niches) can be scaled up to regimes and landscapes building on transition theory. In assessing how niches can become regimes and landscapes, we look at the different concepts discussed in this chapter—resource curse, resource nationalism, dependency, the politics of ideas, socio-environmental justice, impact investing and so on. In addition, we assess the extent to which initiatives towards LFFU transitions in different countries and globally are linked to the socio-relational and ecological-relational dimensions of inclusive development (see Figure 2.3).

2.5. Conclusion

This chapter has created a theoretical background and analytical framework for this book. It uses an inclusive development lens which is operationalised to assess the challenge of LFFU and stranded assets. It examines the theories on the role of different actors (social movements, business and investors, and states) in promoting LFFU, the kinds of motivations they have, the arguments they use and the approaches they develop to promote or resist LFFU. It then shows how individual ideas (niches) can be scaled up to regimes and landscapes building on transition theory.

Finally, let us sum up what we will do in the rest of this book. We will (a) focus on actors, arguments and approaches; (b) in assessing actors—we will also look at their motivations and how they have been able to use their power to adopt certain arguments and approaches that enable LFFU; (c) we will assess how individual activities (niches) can be scaled up to regimes and landscapes thus giving our analysis a dynamic approach; (d) we will examine the trade-offs and synergies between relevant SDGs (see Table 1.1) and (e) we will assess how and under what conditions the transitions can be socio-ecologically inclusive and just.

References

Aarhus Convention on Access to Information, Public Participation in Decision-making and Access to Justice in Environmental Matters. (1998). UNECE [United Nations Economic Commission for Europe].

Abbott, K. W., & Snidal, D. (2009). The governance triangle: Regulatory standards institutions and the shadow of the state. In W. Mattli & N. Woods (Eds.), *The politics of global regulation* (pp. 44–88), Princeton University Press. https://doi.org/10.1515/9781400830732.44

Acselrad, H. (2010). Ambientalização das lutas sociais—o caso do movimento por justiça ambiental. *Estudos Avançados, 24*(68), 103–119. https://doi.org/10.1590/S0103-40142010000100010

Ansar, A., Caldecott, B., & Tilbury, J. (2013). *Stranded assets and the fossil fuel divestment campaign: What does divestment mean for the valuation of fossil fuel assets?* Smith School of Enterprise and the Environment, University of Oxford. https://www.smithschool.ox.ac.uk/publications/reports/SAP-divestment-report-final.pdf

Bebbington, A. (2013). Natural resource extraction and the possibilities of inclusive development: Politics across space and time. Working Paper 21. Effective States and Inclusive Development Research Centre (ESID). https://www.effective-states.org/working-paper-21/

Bebbington, A., Abdulai, A. G., Humphreys Bebbington, D., Hinfelaar, M., & Sanborn, C. (2018a). *Governing extractive industries: Politics, histories, ideas.* Oxford University Press. https://doi.org/10.1093/oso/9780198820932.001.0001

Bond, P. (2013). Justice. In C. Death (Ed.), *Critical environmental politics* (pp. 133–145). Routledge

Bond, P. (2018). Social movements for climate justice during the decline of global governance from international NGOs to local communities. In L. Sharachchandra, E. S. Brondizio, J. Byrne, G. M. Mace & J. Martinez-Alier (Eds.), *Rethinking environmentalism: Linking justice, sustainability, and diversity* (pp. 153–182). MIT Press. https://doi.org/10.7551/mitpress/11961.001.0001

Bridge, G. (2008). Global production networks and the extractive sector: Governing resource-based development. *Journal of Economic Geography, 8*(3), 389–419. https://doi.org/10.1093/jeg/lbn009

Bugg-Levine, A., & Emerson, J. (2011). Impact investing: Transforming how we make money while making a difference. *Innovations: Technology, Governance, Globalization, 6*(3), 9–18. https://doi.org/10.1162/INOV_a_00077

Buttel, F. H. (1992). Environmentalization: Origins, processes, and implications for rural social change. *Rural Sociology, 57*(1), 1–27. https://doi.org/10.1111/j.1549-0831.1992.tb00454.x

Calhoun, C. (1993). "New social movements" of the early nineteenth century. *Social Science History, 17*(3), 385–427. https://doi.org/10.1017/S0145553200018642

Campbell, J. L. (1998). Institutional analysis and the role of ideas in political economy. *Theory and Society, 27*(3), 377–409.

Carbon Tracker Initiative. (N.d.). Terms list. https://carbontracker.org/resources/terms-list/#carbon-budgets

Clarkin, J. E., & Cangioni, C. L. (2016). Impact investing: A primer and review of the literature. *Entrepreneurship Research Journal, 6*(2), 135–173. https://doi.org/10.1515/erj-2014-0011

Coronil, F. (2008). It's the oil, stupid! *Revista: Harvard Review of Latin America, 8*(1), 19–20. https://revista.drclas.harvard.edu/its-the-oil-stupid-an-overview/

De Fazio, G. (2012). Legal opportunity structure and social movement strategy in Northern Ireland and southern United States. *International Journal of Comparative Sociology, 53*(1), 3–22. https://doi.org/10.1177/0020715212439311

De Haan, J., & Rotmans, J. (2011). Patterns in transitions: Understanding complex chains of change. *Technological Forecasting and Social Change, 78*(1), 90–102. https://doi.org/10.1016/j.techfore.2010.10.008

Diani, M., & Bison, I. (2004). Organizations, coalitions, and movements. *Theory and Society, 33*, 281–309. https://doi.org/10.1023/B:RYSO.0000038610.00045.07

Díaz, S., et al. (2018). Assessing nature's contributions to people. *Science, 359*(6373), 270–272. https://doi.org/10.1126/science.aap8826

Dinerstein, A. (2014). *The politics of autonomy in Latin America: The art of organising hope.* Palgrave Macmillan

Ebbesson, J. (2007). Public participation and privatisation in environmental matters: An assessment of the Aarhus Convention. In D. Bodansky, J. Brunee & E. Hey (Eds.), *The Oxford handbook of international environmental law.* Oxford University Press.

Edelman, M. (2001). Social movements: Changing paradigms and forms of politics. *Annual Review of Anthropology, 30*(1), 285–317. https://doi.org/10.1146/annurev.anthro.30.1.285

Escazú Agreement. (2018). Regional Agreement on Access to Information, Public Participation and Justice in Environmental Matters in Latin America and the Caribbean, United Nations (adopted 4 March 2018, entered into force on 22 April 2021). https://repositorio.cepal.org/bitstream/handle/11362/43583/1/S1800428_en.pdf

Escobar, A. (2006). Difference and conflict in the struggle over natural resources: A political ecology framework. *Development, 49*, 6–13. https://doi.org/10.1057/palgrave.development.1100267

Fairhead, J., Leach, M., & Scoones, I. (2012). Green grabbing: A new appropriation of nature? *Journal of Peasant Studies, 39*(2), 237–261.

Fuchs, D. (2013). Theorizing the power of global companies. In J. Mikler (Ed.), *The handbook of global companies* (pp. 77–95). John Wiley & Sons.

Geels, F. W. (2002). Technological transitions as evolutionary reconfiguration processes: A multi-level perspective and a case-study. *Research Policy, 31*(8–9), 1257–1274. https://doi.org/10.1016/S0048-7333(02)00062-8

Geels, F. W. (2018). Disruption and low-carbon system transformation: Progress and new challenges in sociotechnical transitions research and the multi-level perspective. *Energy Research & Social Science, 37*, 224–231.

Geels, F. W., Berkhout, F., & Van Vuuren, D. (2016). Bridging analytical approaches for low-carbon transitions. *Nature Climate Change, 6*, 576–583. https://doi.org/10.1038/nclimate2980

Geels, F. W., & Schot, J. (2007). Typology of sociotechnical transition pathways. *Research Policy, 36*(3), 399–417. https://doi.org/10.1016/j.respol.2007.01.003

Griffin, P. A., Jaffe, A. M., Lont, D. H., & Dominguez-Faus, R. (2015). Science and the stock market: Investors' recognition of unburnable carbon. *Energy Economics, 52(Part A)*, 1–12. https://doi.org/10.1016/j.eneco.2015.08.028

Gupta, J., Pouw, N. R. M., & Ros-Tonen, M. A. F. (2015). Towards an elaborated theory of inclusive development. *European Journal of Development Research, 27*(4), 541–559. https://doi.org/10.1057/ejdr.2015.30

Gupta, J., & Lebel, L. (2020). Access and allocation in earth system governance: Lessons learnt in the context of the Sustainable Development Goals. *International Environmental Agreements: Politics, Law and Economics, 20*, 393–410. https://doi.org/10.1007/s10784-020-09486-4

Harriss-White, B. (2006). Poverty and capitalism. *Economic and Political Weekly, 41*(13), 1241–1246.

Hickey, S., Sen, K., & Bukenya, B. (2014). *The politics of inclusive development: Interrogating the evidence*. Oxford University Press. https://doi.org/10.1093/acprof:oso/9780198722564.001.0001

Hilson, G. (2002). Small-scale mining and its socio-economic impact in developing countries. *Natural Resources Forum, 26*, 3–13. https://doi.org/10.1111/1477-8947.00002

Höchstädter, A. K., & Scheck, B. (2014). What's in a name: An analysis of impact investing understandings by academics and practitioners. *Journal of Business Ethics, 132*, 449–475 (2015). https://doi.org/10.1007/s10551-014-2327-0

Hogenboom, B. (2012). Depoliticized and repoliticized minerals in Latin America. *Journal of Developing Societies, 28*(2), 133–158. https://doi.org/10.1177/0169796X12448755

Hogenboom, B. (2018). Latin America and China. In J. Cupples, M. Palomino-Schalscha & M. Prieto (Eds.), *Routledge handbook of Latin American development* (pp. 179–191). Routledge.

Karl, T. L. (1997). *The paradox of plenty: Oil booms and petro-states*. University of California Press.

Kaul, I., Conceicao, P., Le Goulven, K., & Mendoza, R. U. (Eds.). (2004). *Providing global public goods: Managing globalization*. Oxford University Press.

Kemp, R., & Loorbach, D. (2003). Governance for sustainability through transition management. Paper presented at the EAEPE 2003 Conference, 7–10 November 2003, Maastricht, the Netherlands.

Kramer, R. C. (2013). Carbon in the atmosphere and power in America: Climate change as state–corporate crime. *Journal of Crime and Justice, 36*(2), 153–170. https://doi.org/10.1080/0735648X.2012.752252

Larana, E., Johnston, H., & Gusfield, J. R. (2009). *New social movements: From ideology to identity*. Temple University Press.

Linnenluecke, M. K., Meath, C., Rekker, S., Sidhu, B. K., & Smith, T. (2015). Divestment from fossil fuel companies: Confluence between policy and strategic viewpoints. *Australian Journal of Management, 40*(3), 478–487. https://doi.org/10.1177/0312896215569794

Maduro, M., Pasi, G., & Misuraca, M. (2018). *Social impact investment in the EU: Financing strategies and outcome oriented approaches for social policy innovation: Narratives, experiences, and recommendations*. Publications Office of the European Union.

Martinez-Alier, J., Baud, M., & Sejenovich, H. (2016). Origins and perspectives of Latin American environmentalism. In F. de Castro, B. Hogenboom & M. Baud (Eds.), *Environmental Governance in Latin America* (pp. 29–57). Palgrave Macmillan.

Matikainen, S., & Soubeyran, E. (2022, July 27). *What are stranded assets?* Grantham Research Institute on Climate Change and the Environment. https://www.lse.ac.uk/granthaminstitute/explainers/what-are-stranded-assets/

McAdam, D., Tarrow, S., & Tilly, C. (2001). *Dynamics of contention*. Cambridge University Press. https://doi.org/10.1017/CBO9780511805431

McCarthy, J., & Zald, M. (1977). Resource mobilization and social movements: A partial theory. *American Journal of Sociology, 82*(6), 1212–1241. http://www.jstor.org/stable/2777934

McKibben, B. (2012, July 19). Global warming's terrifying new math: Three simple numbers that add up to global catastrophe—and that make clear who the real enemy is. *Rolling Stone*. https://www.rollingstone.com/politics/politics-news/global-warmings-terrifying-new-math-188550/

Meadowcroft, J. (2009). What about the politics? Sustainable development, transition management, and long term energy transitions. *Policy Sciences, 42*, 323–340. https://doi.org/10.1007/s11077-009-9097-z

Melucci, A. (1996). Youth, time and social movements. *YOUNG, 4*(2), 3–14. https://doi.org/10.1177/110330889600400202

Mosse, D. (2010). A relational approach to durable poverty, inequality and power. *Journal of Development Studies, 46*(7), 1156–1178. https://doi.org/10.1080/00220388.2010.487095

Nicholls, A. (2010). The legitimacy of social entrepreneurship: Reflexive isomorphism in a pre–paradigmatic field. *Entrepreneurship Theory and Practice, 34*(4), 611–633. https://doi.org/10.1111/j.1540-6520.2010.00397.x

Nicholls, A., & Pharoah, C. (2008). *The landscape of social investment: A holistic topology of opportunities and challenges.* Skoll Centre for Social Entrepreneurship.

O'Rourke, D., & Connolly, S. (2003). Just oil? The distribution of environmental and social impacts of oil production and consumption. *Annual Review of Environment and Resources, 28*(1), 587–617. https://doi.org/10.1146/annurev.energy.28.050302.105617

Peters, S. (2017). Beyond curse and blessing: Rentier society in Venezuela. In B. Engels & K. Dietz (Eds.), *Contested extractivism, society and the state: Struggles over mining and land* (pp. 45–68). Palgrave Macmillan.

Rammelt, C. F., & Gupta, J. (2021). Inclusive is not an adjective, it transforms development: A post-growth interpretation of inclusive development. *Environmental Science & Policy, 124*, 144–155. https://doi.org/10.1016/j.envsci.2021.06.012

Rauniyar, G., & Kanbur, R. (2009). Inclusive growth and inclusive development: A review and synthesis of Asian Development Bank literature. *Journal of the Asia Pacific Economy, 15*(4), 455–469. https://doi.org/10.1080/13547860.2010.517680

Rempel, A., & Gupta, J. (2020). Conflicting commitments? Examining pension funds, fossil fuel assets and climate policy in the Organisation for Economic Co-operation and Development (OECD). *Energy Research & Social Science, 69*, 101736. https://doi.org/10.1016/j.erss.2020.101736

Ross, M. L. (1999). The political economy of the resource curse. *World Politics, 51*(2), 297–322.

Rotmans, J., Kemp, R., & Van Asselt, M. (2001). More evolution than revolution: Transition management in public policy. *Foresight, 3*(1), 15–31. https://doi.org/10.1108/14636680110803003

Rozenberg, J., Davis, S. J., Narloch, U., & Hallegatte, S. (2015). Climate constraints on the carbon intensity of economic growth. *Environmental Research Letters, 10*(9). https://doi.org/10.1088/1748-9326/10/9/095006

Sardy, M., & Lewin, R. (2016). Towards a global framework for impact investing. *Academy of Economics and Finance Journal, 7*, 73–80. http://www.economics-finance.org/aefj/issues/AEFJ_volume7.pdf

Setzer, J., & Benjamin, L. (2019). Climate litigation in the Global South: Constraints and innovations. *Transnational Environmental Law, 9*(1), 77–101. https://doi.org/10.1017/s2047102519000268

Setzer, J., & Vanhala, L. C. (2019). Climate change litigation: A review of research on courts and litigants in climate governance. *WIREs Climate Change, 10*(3), e580. https://doi.org/10.1002/wcc.580

Snow, D. A., Soule, S. A., Kriesi, H., & McCammon, H. J. (2019). Introduction: Mapping and opening up the terrain. In D. A. Snow, S. A. Soule, H. Kriesi & H. J. McCammon (Eds.), *The Wiley Blackwell companion to social movements* (2nd ed.; pp. 1–16). John Wiley & Sons. https://doi.org/10.1002/9781119168577

Sovacool, B. K. (2016). The political ecology and justice of energy. In T. Van de Graaf, B. K. Sovacool, A. Ghosh, F. Kern & M. T. Klare (Eds.), *The Palgrave handbook of the international political economy of energy* (pp. 529–558). Palgrave Macmillan.

Stec, S., & Jendrośka, J. (2019). The Escazú Agreement and the regional approach to Rio Principle 10: Process, innovation, and shortcomings. *Journal of Environmental Law, 31*(3), 533–545. https://doi.org/10.1093/jel/eqz027

Strauss, A., Corbin, J. M., & Corbin, J. (1998). *Basics of qualitative research: Techniques and procedures for developing grounded theory.* Sage.

Svampa, M. (2015). Commodities consensus: Neoextractivism and enclosure of the commons in Latin America. *South Atlantic Quarterly, 114*(1), 65–82. https://doi.org/10.1215/00382876-2831290

Temper, L., Del Bene, D., & Martinez-Alier, J. (2015). Mapping the frontiers and front lines of global environmental justice: the EJAtlas. *Journal of Political Ecology, 22*(1), 255–278. https://doi.org/10.2458/v22i1.21108

Tilly, C. (1978). *From mobilization to revolution.* Addison-Wesley.

UNCED [United Nations Conference on Environment and Development]. (1992). Rio Declaration on Environment and Development. A/CONF.151/26 (Vol. I). https://www.un.org/en/development/desa/population/migration/generalassembly/docs/globalcompact/A_CONF.151_26_Vol.I_Declaration.pdf

UNDP [United Nations Development Programme]. (1994). *Human development report, 1994.* http://hdr.undp.org/sites/default/files/reports/255/hdr_1994_en_complete_nostats.pdf

UNEP [United Nations Environment Programme]. (2018). *Inclusive wealth report.*

UNGA [United Nations General Assembly]. (2015, September 25). *Transforming our world: The 2030 Agenda for Sustainable Development.* A/RES/70/1.

Vanhala, L. (2012). Legal opportunity structures and the paradox of legal mobilization by the environmental movement in the UK. *Law & Society Review, 46*(3), 523–556. https://doi.org/10.1111/j.1540-5893.2012.00505.x

Vanhala, L. (2018). Is legal mobilization for the birds? Legal opportunity structures and environmental nongovernmental organizations in the United Kingdom, France, Finland, and Italy. *Comparative Political Studies, 51*(3), 380–412. https://doi.org/10.1177/0010414017710257

Van Norren, D. E. (2017). *Development as service: A happiness,* ubuntu *and* buen vivir *interdisciplinary view of the SDGs*. PhD Thesis, Tilburg University.

Walker, G. (2012). *Environmental justice: Concepts, evidence and politics.* Routledge.

Watts, M. (2000). Political ecology. In E. Sheppard & T. J. Barnes (Eds.), *A companion to economic geography* (pp. 257–274). Blackwell

Watts, M. J. (2005). Righteous oil? Human rights, the oil complex, and corporate social responsibility. *Annual Review of Environment and Resources, 30*, 373–407. https://doi.org/10.1146/annurev.energy.30.050504.144456

WCED [World Commission on Environment and Development]. (1987). *Our common future,* Oxford University Press.

Wilson, B. M., & Cordero, J. C. R. (2006). Legal opportunity structures and social movements: The effects of institutional change on Costa Rican politics. *Comparative Political Studies, 39*(3), 325–331.

Zald, M. (2000). Ideologically structured action: An enlarged agenda for social movement research. *Mobilization, 5*(1), 1–16. https://doi.org/10.17813/maiq.5.1.h8ru112776715vq7

Zoomers, A. (2010). Globalisation and the foreignisation of space: Seven processes driving the current global land grab. *Journal of Peasant Studies, 37*(2), 429–447. https://doi.org/10.1080/03066151003595325

3. Socio-environmental Movements and LFFU: Framing, Tactics and Court Cases

Joyeeta Gupta, Arthur Rempel, Isabelle Hillson, Laura Johnson, Alfie Begley and Nina Brander

Abstract

This chapter examines how actors such as socio-environmental movements can promote through a variety of tactics a niche change in one area which can possibly ripple into a regime or landscape change. The chapter presents a number of fossil fuel-oriented social movements and identifies lessons learnt from criteria for successful social movements. Successful movements operated in politically open states, focused on states and companies, mobilized grievances, had clear local and economic frames, used lobbying and court cases. There are hundreds of court cases globally, but in the Global South these are just beginning. Most court cases led to decisions that could enhance mitigative action and are more successful than the other strategies of social movements. Courts flourish better in more democratic countries, however, so far they have tended to be niche ideas or sometimes regime-change events.

Keywords: fossil fuels, climate change, social movements, court cases, success stories

3.1. Introduction

A key bottom-up actor is a social movement. Local to global socio-environmental movements are growing, protesting injustices and socio-ecological exclusion. Many of these movements focus on energy-related conflicts, like

---

Gupta, J., Hogenboom, B., Rempel, A. & Olofsson, M. (eds), *Leaving Fossil Fuels Underground: Actors, Arguments and Approaches in the Global South and Global North.*
Taylor & Francis Group, 2024
doi: 10.5117/9789048560370_CH03

coal mining, oil drilling or financial incoherence. The key gaps in knowledge include: the mechanisms with which socio-environmental movements successfully drive change at the niche (e.g. stop or promote an activity), regime (e.g. changes in precedent with further impacts) and landscape (e.g. changes in policy) levels; the characteristics of successful campaigns; and why and how they are increasingly using the court system, and how such use of the courts can actually enable a niche idea to become a regime or even landscape idea. Although we take a global perspective, we focus on the Global South and explore whether we can identify key trends and assess relationships with the SDGs.

Against this background, this chapter addresses the following questions: How are socio-environmental movements worldwide addressing the challenge of leaving fossil fuels underground (LFFU), with a particular focus on their role in changing legislation and using the court system? This chapter first builds on the theoretical elaborations in Chapter 2 (see 3.2), presents an overview of environmental movements to LFFU worldwide and lessons learnt (see 3.3), discusses how socio-environmental movements use courts to promote change (see 3.4), before drawing conclusions (see 3.5).

3.2. Building on the analytical framework

Chapter 2 presented an inclusive development approach to the energy transition and discussed theoretical approaches on socio-environmental movements. We now look at how socio-environmental movements can promote a *niche* change in one area which can possibly ripple into a *regime* or *landscape* change. A regime change could be, for example, when a legal precedent that derives from a court case has impact on other similar court cases. A landscape change is when the social movement is also able to change regional, national or even global policy.

Despite the growing urgency of addressing the climate emergency, local to global governments appear to be reacting too slowly. This has spurred a reaction from socio-environmental movements, which are increasingly collaborating to oppose fossil fuel extraction, production and use (Piggot, 2018). These movements are "glocal" (local + global) in nature and have a strong counter-hegemonic role (Giugni, 1999). They may focus on discursive issues—e.g. *buen vivir* and post-extractivism in Latin America; *ubuntu* and eco-feminism in Africa; human rights and well-being in India; and the degrowth and Occupy movements in the Global North. These discursive issues aim to "reconstruct the humanity–nature relation along truly sustainable lines that place human flourishing and grassroots democratic control at the centre" (Carroll & Ratner, 2010, p. 20) and challenge existing knowledge systems by emphasising "ancient worldviews with current relevance, or new frameworks and visions that present systemic alternatives for human and planetary well-being," or push for the "decolonisation of knowledge systems and epistemologies" (Demaria & Kothari, 2017, p. 2589). Increasingly, socio-environmental movements are using scientific agendas to justify their action. They may also be more pragmatic and focus on specific principles and instruments that they think are inappropriate and demand change.

Moreover, socio-environmental movements in relation to LFFU focus either on the lack of environmental commitment in policies or action, on the unjust way in which such policies are crafted either at local level (Bond, 2018), or the lack of credible implementation of the common but differentiated responsibilities and respective capabilities principles at the global level (Bos & Gupta, 2016).

As stated in Chapter 2 and illustrated in Figure 3.1, socio-ecological movements aim to: (a) change policy and precedent through lobbying with legislators, advocacy with administrators and litigation in courts (Piggot, 2018, p. 946); (b) educate the public through information campaigns, media campaigns and stunts, and campaigns to change social norms and attitudes by the stigmatisation of the fossil fuel industry; and (c) more direct action that can range from boycotts of consumer products to violent tactics such as criminal activities which may physically harm people and property (Chenoweth & Stephan, 2011) or physically stopping extraction by stopping construction and/or blocking access to fossil fuel infrastructure or its development, particularly through the use of "soft blockades" (Piggot, 2018, p. 946; Bond, 2018). Protestors often risk repression and violence, and in 10% of the cases assessed are even assassinated (Temper et al., 2020); the stakes are clearly skyrocketing. Despite personal risks, local movements are increasingly successful in ensuring that at least a quarter of fossil fuel projects opposed are suspended or delayed (Temper et al., 2020). Protests have not only focused on fossil fuel, but also other low carbon energy (e.g. hydropower, biomass, renewable energy) projects, where roughly one-third of these face high-intensity conflicts. About half of such projects involve Indigenous rights (Temper et al., 2020) where specific groups of people are forced to sacrifice their rights and where local people are forced to sacrifice their well-being for the energy needs of society—creating unequal risks and benefits (Tramel, 2016; Borras & Franco, 2018). Social movements have focused on a combination of social (participation, racism, Indigenous exploitation) and ecological (energy justice, climate justice) issues.

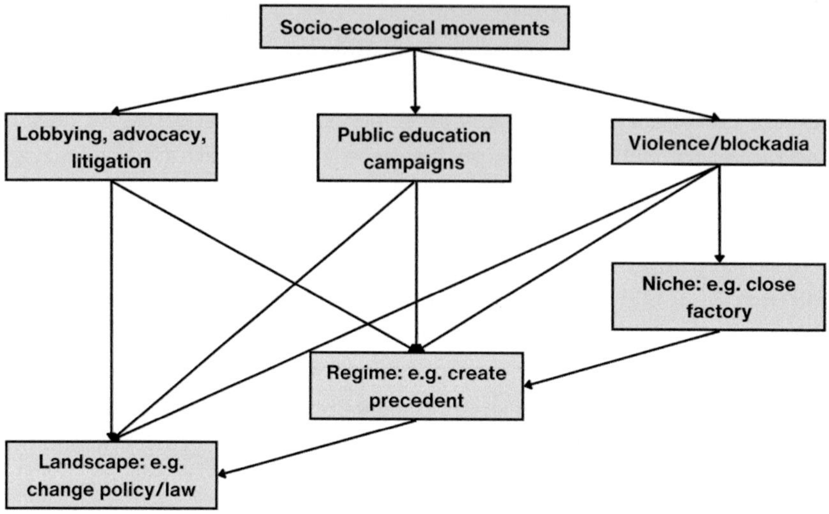

Figure 3.1. How socio-ecological movements influence niches, regimes and landscapes.

We focus here on campaigns to change policy and precedent predominantly through lobbying, advocacy and litigation. Socio-environmental movements can use advocacy to change policies or lobbying to change the way legislation is developed, but if either of these approaches fail (for example, because of competing vested interests), they can use the judiciary. Research has shown that rich lobbyists have huge influence on legislative processes (Presthus, 1974), for example, the five largest fossil fuel companies use $200 million annually to put pressure on governments, spending $2 million on social media ads in the US alone to promote fossil fuel production (Laville, 2019). Social movements can scarcely compete with such lobbying processes and instead often use the court system. Using the court system is strategic and potentially effective because it sets a precedent that can have an impact on future such cases (regime change), and in the best-case scenario can lead to a change in national laws and policies (landscape change). Hence, since 2000, there have been a growing number of court cases on climate change that are relevant for LFFU (Gupta, 2007; Gupta, 2014; Peel & Osofsky, 2015; Martine & Alves, 2019). However, there are also increasing numbers of SLAPP (strategic lawsuit against public participation) court cases where companies are suing NGOs and others in such a way that they become afraid, or financially unable, to use the court system.

While policy is normally made by the legislative and executive arms of government, especially in common law countries, courts will entertain a lawsuit if they feel that the legislature and executive are failing to protect the rights of the people. In civil law countries, the courts are more reluctant to intervene, but even here there is a growing trend of cases. Recognising that "the courtroom doors are always open" (Peel & Osofsky, 2015, p. 340) enables socio-environmental movements and non-governmental organisations (NGOs) to bring their complaints into a legal forum. Courts are becoming important players in multilevel climate governance (Peel & Lin, 2019) and since they can also be used to promote the domestic implementation of nationally determined contributions (NDCs) (Rajamani, 2016), scrutinising them becomes vital. The judiciary is technically objective and is not expected to be swayed by power politics. It will look at the legal arguments and see how to bring about an objective judgement on the issues before it. This implies that the courts can be seen as sitting in a happy medium between top-down regulatory processes and bottom-up protests (Li, 2019; Peel & Lin, 2019). However, in many countries, politicians are trying to control the composition of the judiciary, and this affects its independence. Generally, in climate change court cases, there are claimants, defendants and adjudicators. Socio-environmental movements that go to court usually coalesce around an organisation—like an NGO—which are either the litigants or else providing supporting evidence on behalf of the people affected (Peel & Osofsky, 2015; Setzer & Byrnes, 2019). An example includes the recent court case against Shell; here a Dutch appeals court ruled that the parent company is responsible for its Nigerian subsidiary's role in numerous oil leakages in Nigerian farmland—which was spearhead by Milieudefensie (Friends of the Earth Netherlands) (Meijer, 2021).

The ability of socio-environmental movements to use the judicial system depends on the legal opportunity structure (LOS) (Andersen, 2006; Vanhala, 2012; see Chapter 2). The LOS of a country enables social movements to demand justice through the courts. LOS includes the legal stock of substantive norms and rules, the procedural rules determining legal standing or whether someone is eligible to go to court e.g. to represent someone else or a tree/river, etc., and the rules on legal costs (Andersen, 2006; Vanhala, 2012). In some countries (e.g. England), the losing side has to pay the full costs of the process and this can have a "chilling" effect on the plaintiff's willingness to go to court (Vanhala, 2012). Changes in LOS can influence the distribution of power in democratic societies (Wilson & Cordero, 2006). The theory of regulatory outcomes (Peel & Osofsky, 2013) shows that court cases can have direct (changing precedents that change policy rules—i.e. landscape) and indirect outcomes (influencing markets), where the indirect outcomes influence how decisions are taken (Parker & Braithwaite, 2003; Peel & Osofsky, 2013). Indirect influences can be more transformative (Lin, 2012) and lead to game-changing phenomenon (Markell & Ruhl, 2010)—in this context, long-term changes to the existing landscape. There is growing cooperation between actors in the Global North and Global South in developing such court cases—such as collaboration with Dejusticia in South America (Peel & Lin, 2019) or litigants from the Global South using courts in the Global North (Sands, 2016; Jacobs, 2005; Bodansky, 2017), like the ongoing case against French multinational Total S.A. challenging their plans to construct the East African Crude Oil Pipeline, a mega-project set to run over 1,500 km from inland Uganda to the Tanzanian coast; the case has been filed in the French court system by two French NGOs (Friends of the Earth France, Survie) and four Ugandan NGOs (AFIEGO, CRED, Friends of the Earth Uganda, NAVODA).

Our method followed two distinct stages. We first used the EJAtlas (Global Atlas of Environmental Justice) (see 3.3.2) to identify the array of socio-environmental movements that explicitly focus on fossil fuel extraction, transformation or production issues. From this shortlist, we selected "successful" movements for further analysis. For the purposes of this research, "successful" movements are defined as movements that eventually led to policy change and implementation at the national level—i.e. movements that potentially provoked a regime change. We then analyse them further from the perspective of the inclusive energy transition (see Chapter 2) and draw conclusions. In total, 153 documents pertaining to the successful movements were analysed along with 23 semi-structured interviews to determine the key characteristics and conditions under which movements successfully prompted environmental policy change (see 3.3).

Second, we used the Global Climate Change Litigation Database of the Sabin Centre for Climate Change Law (2020) and the Climate Change Laws of the World Database of the Grantham Research Institute on Climate Change and the Environment (2020) to identify 36 relevant cases to analyse that embody both a North–South and energy focus. We subsequently identified an additional 26 cases to include in our analysis through a systematic literature review, and four additional cases after reviewing ECOLEX (2020)—the largest online global environmental law platform—as a potential additional depository for climate litigation. Furthermore, reviewing the actors involved in climate change litigation (CCL) in South Africa, an additional four cases were discovered in the virtual library of the Centre for Environmental Rights in South Africa (2020), and a final case in the Philippines was also uncovered through exploratory discussions with interviewees. These 81 cases were subsequently analysed using the inclusive energy transition theory (see Chapter 2) to explore how litigation and court systems are (successfully) used to drive environmental change and LFFU (see 3.4).

3.3. Environmental movements and climate change: Success stories

3.3.1. Introduction

Environmental movements—movements with a predominantly environmental focus—have covered a wide range of issues, including energy-related conflicts. This section first presents an overview of recent environmental movements covering a broad range of issues and identifies those that had a focus on fossil fuel-related issues (see 3.3.2); we then discuss which of those were successful in prompting policy change and thereby wider regime change (see 3.3.3); and finally we identify key lessons from successful movements in relation to the inclusive energy transition theory (see 3.3.4).

3.3.2. Overview: Environment–energy justice conflicts

The EJAtlas (Global Atlas of Environmental Justice)[1] contains a quasi-exhaustive global inventory of environmental movements that have taken place since 1970. Figure 3.2 presents an overview of the range of issue areas that movements from the EJAtlas focus on.

Movements driven by nuclear energy, fossil fuel energy and climate justice conflicts summed to 588 unique cases (almost 500 targeted fossil energy, with the remaining 90 or so nuclear centred) (see Figure 3.2). Over 20% of all movements have coalesced around energy conflict issues. Mineral ores and building materials (roughly 550 cases), biomass and land conflicts (400 cases) and water management (330 cases) along with fossil energy comprise the top four most popular focuses of these movements. The purple plot in Figure 3.2 also shows that on average, movements from within these four areas were able to stop roughly 15% of the projects that they were contesting.

1 For an elaboration of the EJAtlas and its varied application, see Martinez-Alier (2021), Martinez-Alier et al. (2014), Pérez-Rincón et al. (2019), Scheidel et al. (2020), Temper et al. (2015) and Temper et al. (2018).

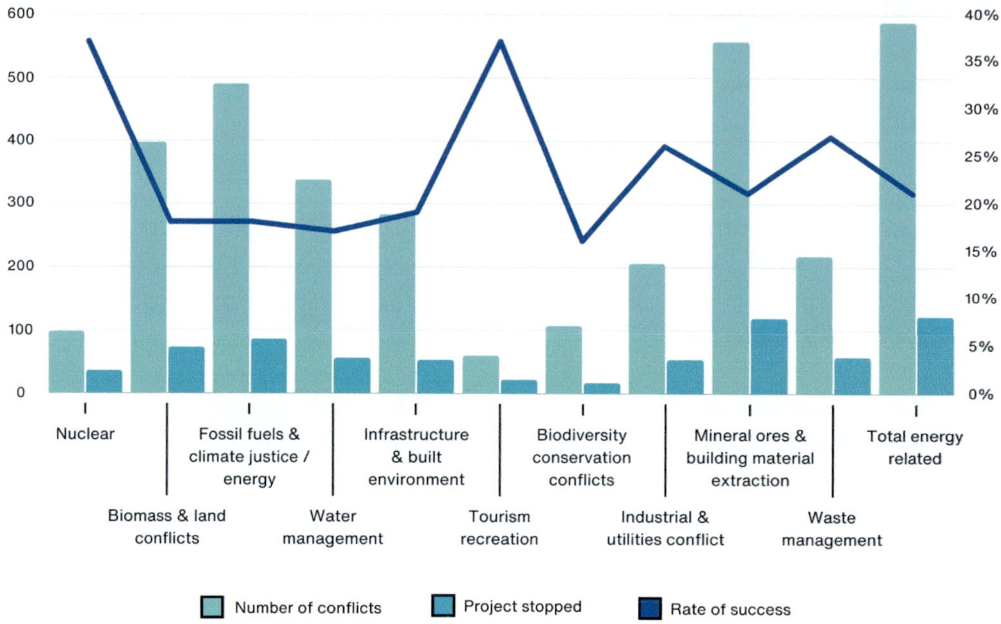

Figure 3.2. Environmental justice conflicts, 1970–2020.
Source: Building on data from EJAtlas.

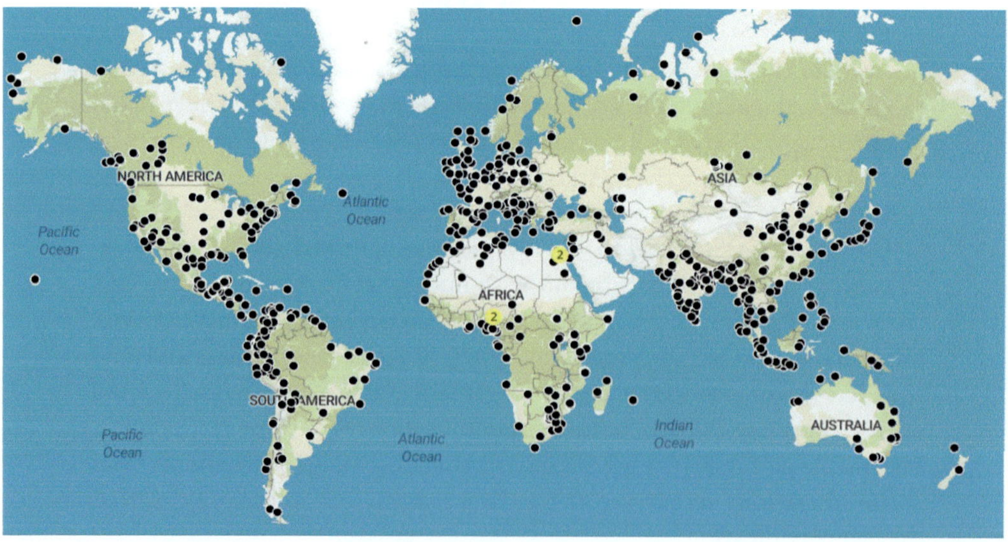

Figure 3.3. Map of energy-related environmental movements.
Source: EJAtlas.

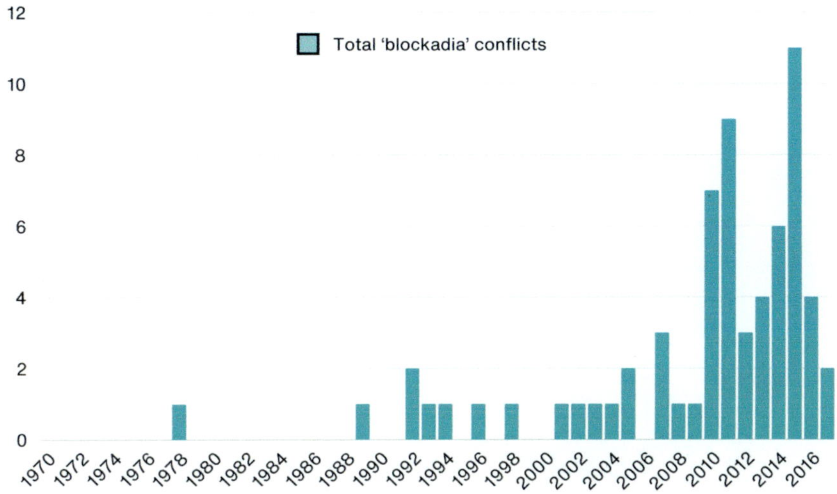

Figure 3.4. Total number of "blockadia" conflicts.
Source: Authors' calculation based on figures from the EJAtlas.

The spatial distribution of energy movements is reflected in Figure 3.3; many are in Latin America, India and Africa but there are also quite a substantial number in the Global North—mainly in Europe and the US. Of these, few were successful in achieving what they had aimed for. Since 1970, 14 environmental movements successfully promoted landscape/regime change: 2 against nuclear, 9 in LFFU, and 3 against water management. Moreover, only 9% of successful energy-focused movements also advocated for and developed alternative renewable energy sources—equivalent to only 2% of all energy-related conflicts. Figure 3.4 shows that many energy-related conflicts escalated to "blockadia" conflicts—a term popularised by Naomi Klein (2015), used to describe conflicts in which protesters "put their bodies on the line" for the movement.

Since 1970, 86 (18% of) movements aspiring to LFFU successfully halted a project and thereby promoted niche level changes; 52 of them were deemed successful by the movements themselves, while the others did not attribute the cessation of the project directly to the movement (see Figure 3.4). When juxtaposed with domestic coal, oil and gas extraction and production rates pertaining to the countries in which these movements to halt fossil fuel projects were situated, we find that net national extraction decreased only in 33% of cases five years after the movement's "success," and thus net extraction increased by 67% of the time (see Figure 3.5). This suggests that environmental movements—particularly targeting the fossil fuel

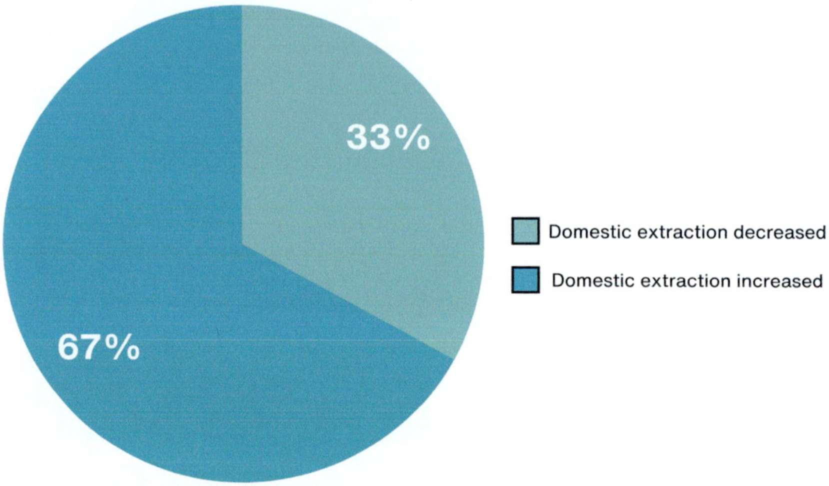

Figure 3.5. Domestic extraction change five years after the fossil fuel project was stopped. Source: Authors' calculation based on figures from the EJAtlas.

industry—struggle to translate these niche level changes into wider regime and landscape-level changes.

Only 10 of these environmental movements that halted a fossil fuel project were able to translate the change at the niche to the regime or landscape level and promoted new legislation influencing the exploration and extraction, transportation and consumption of fossil fuels. These movements were situated both in the Global North and the Global South. The next section explores the intricacies of these movements, drawing key lessons in the context of the Global South and Global North to identify the characteristics that contribute to a truly "successful" environmental movement at the regime and landscape level.

3.3.3. Environmental movements and climate change: Success stories

The 10 successful movements analysed began as early as 1996 (the Mexican maritime space dispute) and ended as late as 2019 (Spain's fracking ban) and lasted for six years on average; four movements were situated in the Global South (all in Latin America) and six in the Global North (one in Europe, two in Australia and three in North America) (see Table 3.1). Of these, only two were able to promote landscape-level change (both of which were in the Global South—Costa Rica and Belize), whereas the other eight (and therefore all six of the movements in the Global North) were only able to prompt regime-level change. The 10 identified movements are briefly contextualised subsequently (see Table 3.1).

In Costa Rica, about a hundred grassroots groups from Limón opposed the 1994 Hydrocarbon Bill, which defined "oil exploration and exploitation as 'public interest'" and mandated the government to grant concessions for oil exploration (Oilwatch Costa Rica, 2005). This ultimately led to a 1999 court case where the Constitutional Court decided that the environment impact assessment (EIA) prepared by Harken Energy was inadequate and subsequently the Supreme Court ruled that concession contracts obtained by Harken were "null and void" (Oilwatch Costa Rica, 2005, p. 5). In 2002, another Harken Energy request for permission to exploit the resource was rejected as it was incompatible with national environmental law, and later a moratorium was enacted that bans oil exploration and extraction. The moratorium was renewed and is now valid through 2050 (Rico, 2019).

In Belize, following an offshore oil discovery in 2008, NGOs (starting with 6 and growing to 41 organisations) began campaigning for an oil moratorium. This coalition united NGOs from various backgrounds—e.g. Citizens Organised for Liberty through Action (COLA), advocating for political voice and rights, and Oceana, pursuing marine ecosystem safety and conservation. In 2007, the Belizean government unanimously passed the Petroleum Operations (Offshore Zone Moratorium) Bill, placing an indefinite moratorium in Belize's waters (Gomez, 2018).

Other successful movements were not able to ensure national level change. In Colombia, Indigenous groups partnered with international actors like Oxfam and successfully objected to oil and coal extraction by Ecopetrol in the Catatumbo River Basin near the Venezuelan border and home to the Moliton Bari Indigenous people. Ecopetrol is a former state-owned company "responsible for administering the nation's hydrocarbon resources," and privatised in the early 2000s (Temper et al., 2013, p. 86). In Mexico, the offshore oilfield Canatrell, owned by state-owned Petróleos Mexicanos (Pemex)—the 10th-largest oil company and responsible for approximately 86% of Mexico's oil production (Quist, 2019)—is located in the biodiversity rich Bay of Campeche with artisanal fishing producing red snappers, prawns, oysters and snooks (Quist, 2019). Local fishermen protested against continued expansion of an exclusion zone by the oil company (Soto et al., 2009) for 13 years before the state decided to reopen an area of 10,000 km², which had been formerly part of the 15,900 km² exclusionary zone (Quist, 2019).

In the Global North, social movements have had limited success. Spain's desire for energy independence (i.e. to reduce its oil and gas imports) (Burgen, 2014; Lin, 2014a, p. 1046) led to a policy to promote fracking and by 2014, 70 permits had been granted and a further 40 were pending (Burgen, 2014). A protest ("Assembly against Hydraulic Fracking") was organised by mostly local NGOs and citizen assemblies and multilateral players campaigned that "Gas Is Not the Solution" (WWF, 2018). In April 2020, the Spanish government announced its intent to ban fracking at the national level through a new law for climate change and transition (Spanish News Today, 2020). Cantabria—with Santander as its capital on the north coast of Spain—was the first autonomous community to bring about new legislation to prohibit fracking.

In Australia, Friends of the Earth (FoE) Melbourne mobilised people through its "Quit Coal" campaign to ban fracking for onshore unconventional gas and successfully acquired a moratorium on conventional gas exploration and drilling in Victoria (Walker, 2016). Victoria was already a heavy coal producing and extracting region; this campaign was part of a broader "No New Fossil Fuels in Victoria" movement, which also prevented several planned coal mines from being constructed (Friends of the Earth Melbourne, 2019). At the same time, the Yes2Renewables (Y2R) movement in Victoria opposed brown coal consumption and advocated for transitioning towards renewable energy sources. Brown coal (the most polluting coal) is extracted from Victoria's Gippsland Basin and used in its power stations (Hughes, 2018, p. 1). A 2014–2016 campaign resulted in a commitment of the state government to a Victorian Renewable Energy Target (VRET) of 25% by 2020 and 40% by 2040 (Ewbank, 2016).

In Canada, the world's fourth-largest oil producer (US Department of Energy, 2019), and Alberta, home to the third-largest global oil reserves in the form of tar sands (Temper et al., 2013, p. 115), Indigenous groups (the Yinka Dene Alliance) and large international NGOs protested against the expansion and by 2016 had successfully prevented the proposed Enbridge Northern Gateway Pipelines (to carry tar sands oil from Alberta to Kitimat) from being built. Canada is also the third-largest global natural gas producer (Ritchie, 2017; EIA, 2019). During the 2012–2013 "gold rush of natural gas" with 18 proposals to transport fracked liquefied natural gas (LNG) from north-eastern British Columbia to the coastal areas in Prince Rupert and Kitimat (Friends of Wild Salmon, n.d.), a public campaign challenged the LNG export facility owned by Petronas, and in 2017 Petronas withdrew its application. This campaign was spearheaded by salmon scientists, climate scientists and "Friends of Wild Salmon" (which includes local businesses and fishers).

In the US, between 2002 and 2012, grassroots and larger environmental NGOs campaigned in Chicago to close the two old and highly polluting coal plants, Fisk and Crawford, and for Chicago to commit to a 100% renewable energy target (Germanos, 2019). This movement organised itself as the

Table 3.1. Contextual information on 10 case studies

Movement	Costa Rica oil moratorium	Belize offshore moratorium	Colombia oil and coal extraction	Mexico maritime space dispute	Cantabria, Spain, fracking ban	Victoria, Australia, UCG fracking ban	Yes 2 Renewables #VRET, Victoria, Australia	Enbridge Northern Gateway Pipelines	Gitwilgyoots LNG export facility	Chicago coal power plants
Country & region	Costa Rica, DC, Global South	Belize, SIDC, Global South	Colombia, DC, Global South	Mexico, DC, Global South	Spain, IC, Global North	Victoria, Australia, IC, Global North	Victoria, Australia, IC, Global North	British Columbia, Canada, IC, G7, Global North	British Columbia, IC, G7, Canada, Global North	Illinois, US, G7, Global North
Energy market: importer/ exporter, domestic fossil fuel production TWhs	Net importer, N/A	Net importer, N/A	Net exporter, Coal: 658.33 Oil: 543.1 Gas: 129	Net exporter, Coal: 72.22 Oil: 1,103.7 Gas: 332.16	Net importer, Coal: 12 N/A	Net exporter, Coal: 3,652.78 Oil: 239.6 Gas: 1,499.61	Net exporter, Coal: 3,652.78 Oil: 239.6 Gas: 1,499.61	Net exporter, Coal: 308.33 Oil: 3,197.1 Gas: 1,691.1	Net exporter, Coal: 308.33 Oil: 3,197.1 Gas: 1691.1	Net exporter, Coal: 3,972.22 Oil: 8,684.12 Gas: 8,996.68
Democracy score	8.13	6.13[25]	7.13	6.09	8.29	9.09	9.09	9.22	9.22	7.96
Relevant elections	2002 General election: small majority for incumbent Social Christian Union Party (centre right)	2015 General election: UDP (centre right) won 3rd term	Governor of Norte Santander Department 2003–2007, went on to run National Hydrocarbons Agency	State Campeche Governor 1997–present, Institutional Revolutionary Party (PRI)	2015 & 2019 Cantabrian elections: People's Party (conservative) & Regionalist Party	2014 state election: Conservatives lost to Labor Party	2014 state election: Conservatives lost to Labor Party	2011–2017 Liberal Party (centre right) in BC state: 2017 BC state election won by NDP & Green Party	2011–2017 Liberal Party (centre right) in BC state: 2017 BC state election won by NDP & Green Party	Chicago mayoral election 2011: Rahm Emmanuel (Democrat) beat incumbent, 2008: Obama implemented pollution regulations
Dates	1999–2003	2010–2017	2002–2006	1996–2017	2011–2019	2011–2016	2013–2016	2010–2016	2015–2017	2002–2012
Commodity	Crude oil	Crude oil	Crude oil, coal	Crude oil	Shale gas fracking	Unconventional gas fracking	Coal, electricity	Crude oil, natural gas, bitumen	Liquid natural gas	Coal
Stage in fossil fuel life cycle	Exploration & extraction	Exploration & extraction	Exploration & extraction	Exploration & extraction	Exploration & extraction	Exploration & extraction	Extraction / alternatives	Transportation	Transportation	Consumption
Success level	National policy change	National policy change	Regional policy change	Regional policy change	Regional policy change	Regional policy change	Regional policy change	Regional policy change	Regional policy change	Regional policy change

Chicago Clean Power Coalition in 2010 growing from 6 to 60 organisations. In doing so, and given the socioeconomic diversity of its members, it also focused on "health, economic and equity issues."

All the above movements were in politically "open" states (The Economist, 2019). Seven out of 10 movements were at the "extractive frontier" opposing the extraction of coal, oil and gas. The other three were in the Global North: two in Canada against the transportation of gas and a movement in Chicago opposed fossil fuel consumption (Scheidel & Schaffartzik, 2019). This discrepancy hints at the "uneven and combined" nature of capitalist development as conflicts in industrialised countries occur further along the fossil fuel life cycle (Scheidel & Schaffartzik, 2019). Seven of the cases used local arguments (local pollution) to mobilise people to protect their own health, using "not in my back yard" (NIMBY) arguments. Economic arguments were utilised in four cases to mobilise support and persuade policymakers. None of the cases emphasised global climate change issues. More often, and particularly in the Global South, local air, water and soil pollution and accompanying ecological and social ramifications were the leading arguments for opposing fossil fuel projects rather than a global climate change narrative. The next section further unpacks the key themes and lessons spanning these 10 movements.

3.3.4. Key common features of successful movements

Frames

We now derive lessons from the successful movements discussed above. The framing of an issue is key to the success of a movement. Structured and robust frames help to mobilise people and groups by amplifying grievances, mobilising resources, capitalising on political opportunity and building on cultural perspectives (Snow, 2013, p. 6).

Six distinct frames were employed by 10 movements, namely; green local environmentalism; economic ideas; climate change; Indigenous rights; health; and conservationism (see Figure 3.6). The "economic" frames mostly centred around employment vis-à-vis tourism and fishing; "green local environmentalism" targets concerns over local air, water and soil pollution and its subsequent impacts on food security and access to drinking water and sanitation; climate change focuses on the global challenges; Indigenous rights focus on the way in which their lands and waters were damaged; health focuses on the health impacts of the environmental consequences; and conservationism focuses on the need to conserve and protect nature.

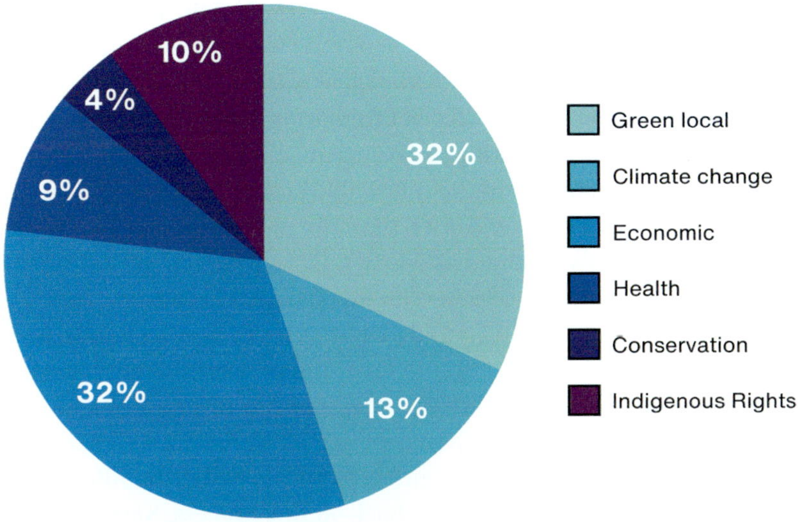

Figure 3.6. Dominant frames of successful movements on LFFU.
Source: Authors' calculation based on figures from the EJAtlas.

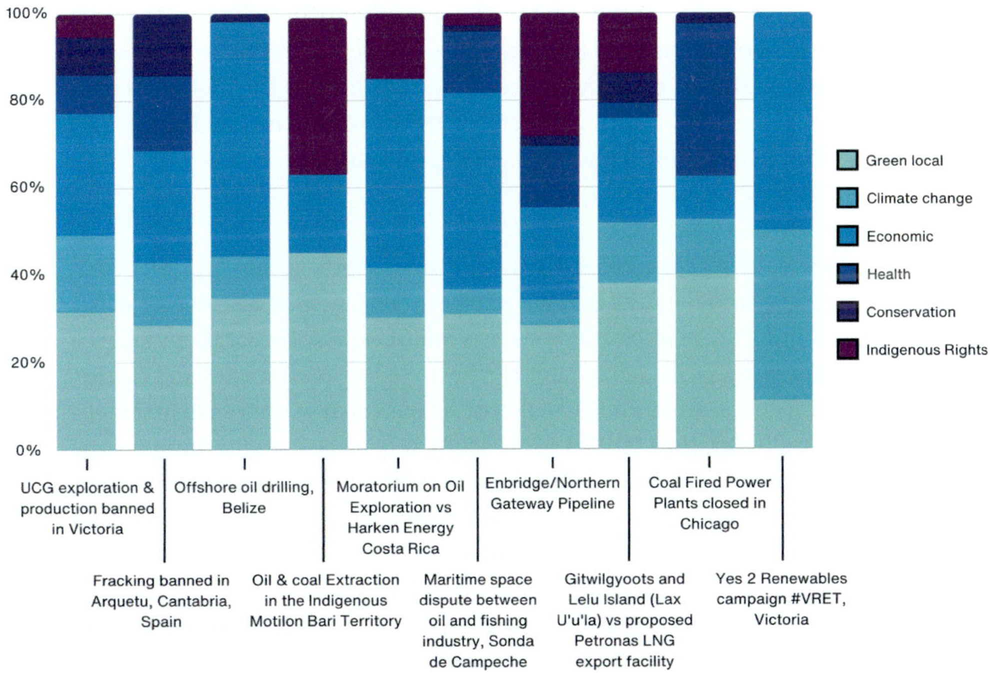

Figure 3.7. Breakdown of the key frames used per movement.
Source: Authors' calculation based on figures from the EJAtlas.

Figure 3.7 demonstrates the breakdown of frames within the 10 movements, illustrating that all included economic frames and arguments to some extent (illustrated in purple), although it was the overpowering frame for a subset; the dispute in Mexico centred around fishing, in Belize and Costa Rica it focused on fishing and tourism, and the Y2R campaign in Victoria centred on jobs lost in the renewable energy sector. A local frame was also consistently present and often the main frame. Two cases concerned fracking—in Spain also framed as NIMBY. Once broad mobilisation was achieved, both campaigns tied in climate change. Three movements on Indigenous territory, two in Canada against fossil fuel transportation and one in Colombia used frames of local impacts on the land, such as detrimental effects on biodiversity, Indigenous food supplies and local economies and culture, rather than centring on Indigenous rights. What is clear with framing is that local arguments are needed to mobilise people and global arguments can then piggyback on this process. Moreover, four movements (Belizean, US and both Canadian movements) strategically selected frames that differed from those needed to mobilise people (i.e. a focus on local arguments) from those needed to persuade policymakers (drawing on economic arguments). The movements in Belize and against the Enbridge pipeline were both mobilised to act in the wake of the 2010 BP Deep Horizon oil spill, but rather than focusing on the global issue of climate change, the campaigns respectively focused on the impacts on Indigenous people and economics (see Figure 3.6).

Tactics

Social movements predominantly target their efforts at the state, although in the case of movements against extraction, they may also target other powerful groups and institutions such as multinational corporations and the media. The success of a movement can be linked to its strategies and tactics, depending on its vulnerability to delegitimisation or openness to influence (Adams & Shriver, 2016; Walker et al., 2008). When the target is the state, the political structure determines tactical repertoires; political opportunities are less frequent in authoritarian states which lack avenues for engagement and repress people often pushing movements to use extra-institutional means (Adams & Shriver, 2016, p. 896; Osa, 2001). Where states have civic space and allow engagement, social movements can use a range of strategies. Second, the more challenging a demand is from the policymaker's perspective, the greater the degree of assertiveness needed by the social movement (Amenta et al., 2010; Skrentny, 2006, p. 1764). Third, all 10 movements used educational outreach programmes (consisting of training and informative events, publishing scientific reports and door-to-door canvassing) to mobilise activists; this enabled awareness building among members of the public and communicating grievances. For instance, the Colombian NGO CODACOP (Corporación de Apoyo a Comunidades Populares) aims to mobilise peasants and Indigenous people through capacity building via training.

Fourth, the three most prominent tactics employed to persuade policy-makers were: lobbying, legal recourse and extra-institutional tactics such as marches and soft blockades. Lobbying (and advocacy) was employed at local, state and national levels, though it was much more prevalent in the former. For instance, the Chicago movement sought to close the two polluting coal facilities (Fisk and Crawford) by passing a city-level policy to impose pollution reductions). Sympathising aldermen of Chicago's city council adopted the movement's plea, and eventually the movement was able to sway enough public support to convince 35 of the 50 council members to support the proposed reductions. Eight movements used lawsuits as a tactic, which was particularly effective for the three movements that involved Indigenous people and rights infringement. For instance, in Costa Rica, the grassroots environmental group ADELA (Acción de Lucha Anti-Petrola) appealed to the Constitutional Court to revoke an exploration concession on the grounds that it violated the International Labour Organization's Convention on Indigenous and Tribal peoples (Oilwatch Costa Rica, 2005, p. 6). Finally, marches and media stunts were used by all movements and soft blockades. For example, local groups in the Australian movement against unconventional gas held public events to attract media attention and declaring themselves "gasfield free"; the protesters locked themselves to the gates and undertook tours of the country. In the anti-Enbridge pipeline campaign, protestors made sure that the pipeline could only be built if the protestors were arrested, increasing the transaction costs for the company and government.

Common features

Common features of the successful movements include that they were all in politically open states which allow protest and focused mostly on the state and sometimes companies. The successful movements were able to bring together large coalitions by "mobilising grievances" and perceptions of injustice. Six of the 10 movements studied actively involved Indigenous communities, although they are only 5% of the global population; this over-representation of Indigenous people reflects their existence at the extractive frontier and exploitation by fossil fuel projects (Gupta & Vegelin, 2016; Harriss-White, 2006). Successful movements had clear frames focusing on local environmental arguments and/or economic arguments (e.g. fishing and tourism unemployment) that were underpinned by scientific reports, but only occasionally linking to climate change, suggesting that "climate change" is in and of itself not an impactful catalyst for environmental movements. Local and economic frames were most successful because they touched on local issues and because local elites may support them (Skrentny, 2006); however, an economic frame is also limited as it inhibits movements from demanding and achieving the structural changes necessary to LFFU. Moreover, lobbying at the local and state government level (all 10 movements) and court cases (8 of 10) were particularly effective while protest marches and soft blockades were the only non-institutional tactics used. Fifth, in order to actually mobilise policy change at national level (Belize and Costa Rica) this was easier as the business interests in fossil fuel were less important than the business interests in tourism and fishing. This also implies that the higher the business interests in fossil fuel the more difficult it is to get policy change (Muttitt & Kartha, 2020, p. 4). Sixth, in terms of temporality and spatiality, the strict definition of success (i.e. policy change, *not* simply halting a fossil project) this study uses only scrutinises about 2% of all environmental justice movements since 1970 (see 3.3.2). However, the above results are probably valid because they cover a long period, covered the Global North and Global South, and resonate with the literature.

3.4. Socio-environmental movements and court cases

3.4.1. Introduction

Eight of the 10 movements in the earlier analysis filed lawsuits and pursued legal action throughout their campaigns (see 3.3.3), 2 of which succeeded in passing policies at the national level that challenged fossil fuel production at the extractive frontier—contributing to efforts to LFFU. This section builds on this finding by exploring documented court cases to study the use of courts by socio-environmental movements in pursuit of LFFU. Many court cases are initiated by NGOs which are not strictly speaking social movements and may not always be demanding "environmental justice." This may also reflect why the database of these court cases may vary from the database of environmental justice movements. Moreover, the database may have an English-language bias and may not have included cases that were not explicitly referred to as "climate change cases."

As of July 2020, the database of the Sabin Centre for Climate Change Law (2020) included 1,576 CCL court cases. A preliminary assessment of these reveals that most cases have been filed in the US (1,214 cases) and the Global North more broadly, creating a "Global North Bias" (Setzer & Vanhala, 2019, p. 12). Box 3.1 provides a short case study of a UK case.

Box 3.1. A recent UK case

A landmark decision in *R (on the application of Friends of the Earth Ltd and others) v. Secretary of State for Business Energy and Industrial Strategy* (2022) results from three separate lawsuits by (1) Friends of the Earth (FoE), (2) Client Earth and (3) the Good Law Project along with Johanna Wheatley against the Secretary of State for Business Energy and Industrial Strategy. The claimants challenged the legality of two policies in October 2021: (1) the Net Zero Strategy and (2) the Heat and Building Strategy, claiming that these violated the UK's Climate Change Act of 2008 (CCA), the Equality Act of 2010 and the Human Rights Act of 1998. The court ultimately ordered that the secretary of state lay before Parliament a fresh report complying with the CCA by March 2023. This case represents a victory for climate justice as the High Court, in this momentous decision, not only upheld the spirit and intent of the CCA, but enforced its provisions as well as the principles of transparency, accountability and equality.

Four waves of scholarship on socio-environmental movements and court cases can be identified. In the first, scholars focused on the range of potential legal arguments that could be used in different jurisdictions (Gupta, 2007) and the scholarship was dominated by legal scholars (Setzer & Vanhala, 2019). Following this, social movements and NGOs started to use the court system, and the second wave developed typologies of litigants, arguments used and the forum or the type of court. These cases quantified emissions and addressed major carbon-emitting corporations (Ganguly et al., 2018). A third wave examined the impact of litigation on society (Peel & Osofsky, 2013). A fourth wave focused on the differences in geographies and the ways in which social movements worldwide learn from each other.

We focus here on the court cases in the Global South. After exploring and presenting an overview of the documented court cases in the Global South (see Figure 3.8), we draw on these four theoretical waves of scholarship in unpacking the predominant actors, objectives and focuses of the sampled CCL cases.

3.4.2. Overview: CCL cases in the Global South

Figure 3.8 presents the global distribution of documented CCL court cases in the Global South, denoting the range of cases per country using a gradient

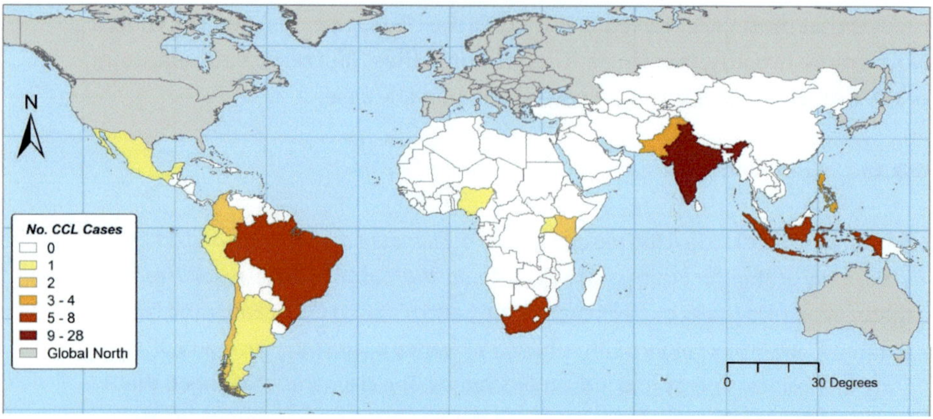

Figure 3.8. Map of climate change cases in the Global South.
Source: Authors' calculation based on figures from the EJAtlas.

scale. For instance, out of our data set, India documented the greatest frequency of CCL cases in the Global South (between 9 and 28), while the next frontrunners (Brazil, South Africa and Indonesia) hosted between 5 and 8 cases each. More generally, CCL court cases were identified in only 15 countries; while it is well emphasised that litigation may be significantly under-reported across many of these jurisdictions (Setzer & Vanhala, 2019; Peel & Lin, 2019), this lack of empirical evidence suggests that at present, CCL is not utilised to its greatest potential as a regulatory tool widely across the Global South. It could also reveal that the database is not complete as our court cases in Belize and Costa Rica are not reported here—implying that there may be an English-speaking bias in the database. Moreover, if court cases focus on local issues, these may not be classified as "climate change cases." Peel and Lin (2019) identified CCL cases in 11 countries in the Global South in 2018; this research indicates a noteworthy geographical expansion, particularly given that the four additional countries in which we identified CCL case claims have been filed in Latin America. This resonates with the earlier analysis of the four successful environmental movements for LFFU from the Global South—all of which were in Latin America (see 3.3.3, and also Chapters 4 and 8)—potentially implying that, while courts are currently not widely used across the entire Global South, actors from specific regions may be recognising its potential in regulating climate action and transnationally disseminating and sharing knowledge and information.

We note that court cases occur more frequently in countries with somewhat stable economies and institutions and that a lack of capacity

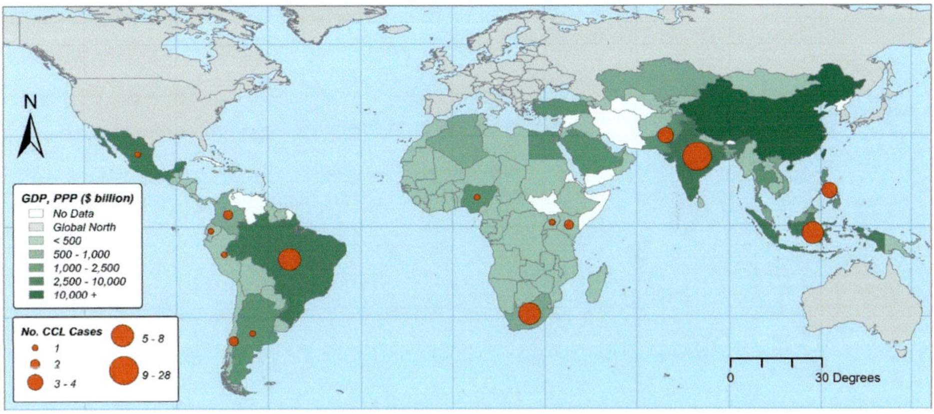

Figure 3.9. Map of climate change cases linked to GDP.
Source: Authors' calculation based on figures from the EJAtlas and World Bank (2020) data.

and the need to focus on immediate concerns may lead local actors not to use these courts (Setzer & Benjamin, 2020; Setzer & Byrnes, 2019). To explore whether this applies throughout the CCL cases in the Global South, Figure 3.9 overlays national GDP (PPP, billions $) data (as a crude estimate of "economic development") onto the distribution of CCL cases from Figure 3.8. Notably, the three leading nations in terms of the pure number of filed CCL cases—India, Brazil and South Africa—are members of the BRICS intergovernmental organisation and are more industrialised than others in the sample—again, purely from a crude, GDP-based angle.

Another factor that may encourage the use of courts is the degree to which a country is democratic and has civic space for local actors. Generally, in democratic countries, courts are seen as a place to "uphold and enforce the law and make the theoretical processes of democracy work more effectively in practice" (Preston, 2016, p. 14; Osofsky, 2005). This could imply that there would be less scope for court cases in authoritarian regimes such as China (Li, 2019, p. 160). Figure 3.9 assesses this by examining whether climate change court cases occur in democracies by matching their distribution. Figure 3.10 show these cases in terms of how the country's rank in terms of the Democracy Index 2019 (EIU, 2020). This superimposition of maps shows that of all the cases, only Chile is identified as a "full democracy," while the other cases have occurred in "flawed democracies" or "hybrid regimes." It is interesting to note that the highest-ranked democracies in the Global South (e.g. Uruguay, Mauritius) had no climate cases in the databases and that no cases appear to have occurred in "authoritarian regimes." This suggest that a democratic society

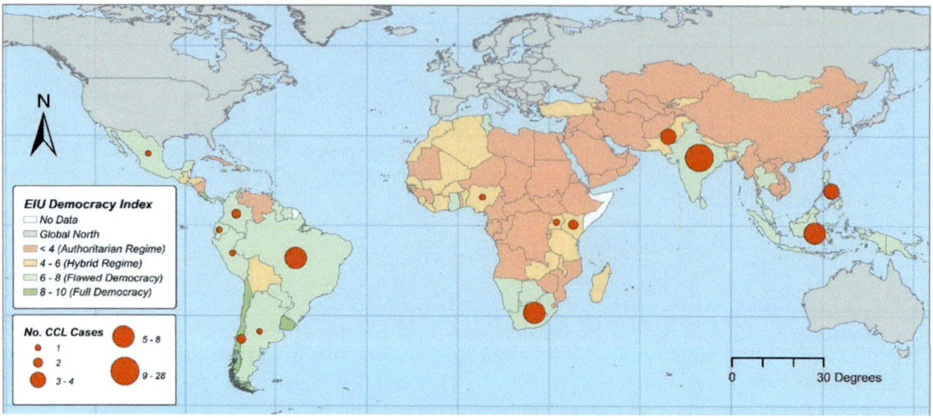

Figure 3.10. Map of climate change cases linked to the Democracy Index 2019.
Source: Authors' calculation based on figures from the EJAtlas and The Economist Intelligence Unit (EIU, 2020).

may be necessary for social actors to use the court system, but that merely being a democracy does not imply that there will be climate change court cases.

Another possible factor that could influence the location of court cases could be the vulnerability of the country concerned or the degree to which it emits greenhouse gases. We have assessed this by overlaying the distribution of cases onto a map with national CO_2 emissions in millions of tonnes (see Figure 3.11), and onto a map that assesses climate vulnerability using the Notre Dame Global Adaptation Initiative Country Index (ND-GAIN, 2019) (see Figure 3.12). This shows that countries with high emissions were also countries with climate litigation (GCA, 2020). India's CO_2 emissions in 2018 accounts for 7% of global emissions (ranking third among the chief global emitters on a total basis, i.e. not a per capita basis) and so it seems logical that 40% of the climate litigation cases studies were in India. This may imply that there is more climate litigation in countries with high emissions.

However, we do not find a similar relationship with respect to vulnerability. Figure 3.12 shows that the most vulnerable countries are not necessarily the ones where there is more litigation. For example, the least vulnerable country in the Global South (Chile, ranking 21st globally) had as many court cases as the most vulnerable (Kenya, ranking 150th globally). Thus, while vulnerability may influence individual actors to use the national courts, it is the mitigation potential that draws these actors to use the courts more (Setzer & Benjamin, 2020; Peel & Lin, 2019).

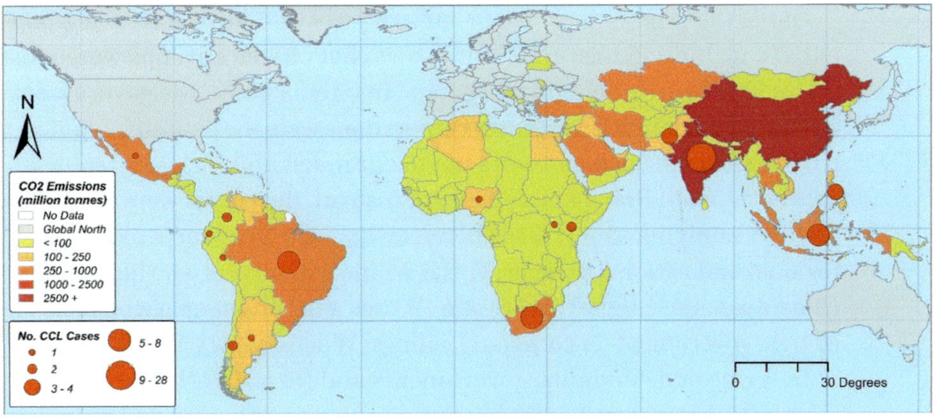

Figure 3.11. Map of climate change cases in relation to total emissions.
Source: Authors' calculation based on figures from the EJAtlas and the GCA (2020).

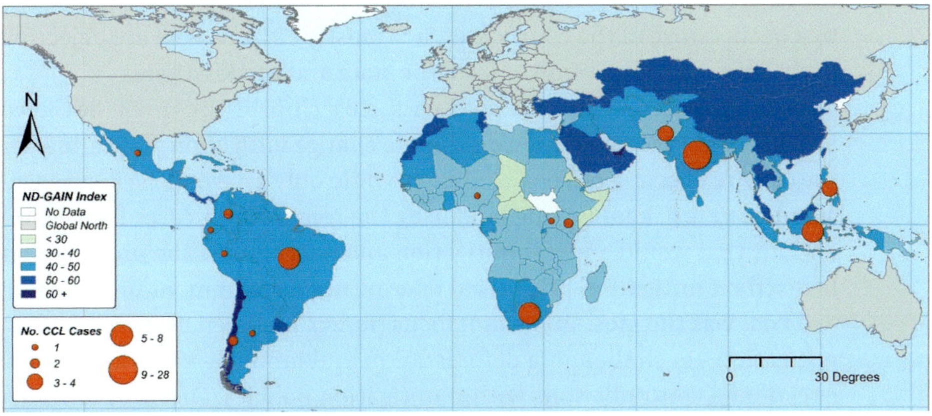

Figure 3.12. Map of climate change cases in relation to vulnerability.
Source: Authors' calculation based on figures from the EJAtlas and the Notre Dame Global Adaptation Initiative (ND-GAIN, 2019).

3.4.3. Key common features of the court cases

In terms of actors (both plaintiffs and defendants) in the court cases in the Global South, we explore both the top-down and bottom-up nature of regulation through the judiciary by adopting a typology from similar studies (Osofsky, 2005; Markell & Ruhl, 2012). We categorise actors into (i) government; (ii) NGO; (ii) individual; and (iv) corporation; along with an additional class (v) court. While courts are part of the judiciary, this was selected to see if courts were willing to act on their own accord with regard to climate change issues.

Out of the 71 cases included in our study, 50 were filed by citizen groups (NGOs: 19; individuals: 40).[2] This shows that citizen's groups were most active in litigating on climate change. In only 14 cases, the governments also acted as plaintiffs; in 4 cases corporations were the plaintiff. Top-down participation in such cases did not occur much and was geographically concentrated in Brazil and Indonesia. Instead, the courts were used from "below" implying that it was citizens who used the courts to challenge regional and national policies. This is in line with others who have studied climate cases and argue that litigation is used as a bottom-up phenomenon to address government (or corporate) failings (Wilensky, 2015; Peel & Lin, 2019).

In terms of defendants, governments and (to a slightly lesser extent) corporations were indisputably the most frequently targeted groups in the climate court cases. About 59 cases focused on the state and 29 on the corporations and sometimes they were both simultaneously the defendant. Only 4 of the 71 cases focused on citizen's groups as the defendant. This shows that at this stage of the court cases, it is mostly citizens who are objecting to the greenhouse gas emissions of the state and corporations.

In terms of arguments and approaches, most climate cases from the Global South (60/71, or 84%) focused on climate change mitigation (e.g. calling for the closure of coal-fired power plants), while only 4 cases (6%) focused on adaptation (e.g. challenging land use); the remaining 7 cases focused on options that would address mitigation and adaptation. The projects that focused on mitigation prioritised energy and development projects (25) and coal was the most important focus (14/25). Eighteen cases focused on mitigation via challenging deforestation and land-use violations and 15 questioned contradictions within mitigation policies. Of the 71 cases, 63 (89%) were proactive in nature, and only 7 were inherently "anti-regulatory" (i.e. reactively challenging existing legislation).[3] This leads us to conclude that the climate change cases in the Global South mostly focus on catalysing positive and proactive change on the climate change mitigation frontier.

Juxtaposing these findings with those of the predominantly involved actor groups indicates that overall, such court cases in the Global South tend to be driven by coalitions of NGOs and/or affected individuals (as plaintiffs) who challenge governments and/or corporations for climate-related violations, the end result of which would very likely yield positive implications for LFFU and thus climate change mitigation. Finally, at the time of research, 57 of the 71 cases had reached a final verdict; 47/57 resulted in "successful" rulings

2 Note that these do not sum to 50, because multiple litigants (and defendants) may exist in any given CCL case

3 The final case was neither proactive nor anti-regulatory

supporting LFFU (in which the court either ruled in favour of the litigants of a proactive CCL case, or against the litigants of the anti-regulatory case), while only 10/57 were deemed unsuccessful by the same metric. Altogether this seems to indicate that in the Global South, courts have thus far offered a promising and effective platform for civil society to mobilise and pursue bottom-up change, synergising with both LFFU and climate change mitigation. Box 3.2 presents some insights from court cases in Africa.

Box 3.2. Court cases in Africa

Most climate relevant court cases in Africa (about 14) have been in five countries. Courts have tried to balance fundamental human rights with socio-economic rights, as fossil fuels are an energy source authorised by existing laws and policies for addressing energy security and poverty alleviation (*Save Lamu and Others v National Environmental Management Authority and Amu Power Co. Ltd.*, 2019; *Earthlife Africa Johannesburg v Minister of Environmental Affairs and Others*, 2017). Most plaintiffs are NGOs on behalf of vulnerable communities and individuals, and most cases occur where there is some civic space and laws do not impede their activities (Poppe & Wolff, 2017; Buyse, 2018). Most defendants are private companies and government agencies that authorise their projects without regard for socio-ecological consequences. NGOs have argued that such projects: (i) violate fundamental human rights (*Gbemre v Shell Petroleum Development Company of Nigeria Limited and Others*, 2005); (ii) have been authorised without (a) a proper investigation of climate change impacts (*Earthlife Africa Johannesburg v Minister of Environmental Affairs and Others*, 2017); (b) proper public participation (*Save Lamu and Others v National Environmental Management Authority and Amu Power Co. Ltd.*, 2019); (c) regard for international environmental law and treaty obligations (*Centre for Food and Adequate Living Rights Limited et al. v Attorney General of the Republic of Uganda et al.*, 2020); and (iii) demonstrate the government's climate inaction (*Mbabazi and Others v The Attorney General et al.*, 2012). Resort to human rights often serves as a "gap filler," in the absence of laws that specifically provide redress for climate change impacts (Savaresi & Auz, 2019). NGOs have sought (i) compensatory damages and (ii) declarations to mitigate climate change impacts. Courts have consistently upheld the requirement for entities to conduct environmental impact assessments properly before the state may issue permits *(Save Lamu and Others v National Environmental Management Authority and Amu Power Co. Ltd., 2019; Earthlife Africa Johannesburg v Minister of Environmental Affairs and Others*, 2017). However, they are not likely to go so far as to forbid fossil fuel investments as long as these are legally permissible.
Source: Mugga et al. (2023).

3.5. Conclusion

This chapter has discussed successful LFFU social movements worldwide and climate change litigation primarily in the Global South (because of the huge coverage in the literature in the Global North). We now link the findings from the socio-environmental movements (see 3.3) and climate change court cases from the Global South (see 3.4) to our broader inclusive energy transition theory (see 2.2.3), by inferring their respective implications for the landscape, regime and niche levels and identifying arguments and approaches spanning social, ecological and relational dimensions of inclusiveness.

First, in order to be able to create a social movement, economic and local framings are needed to convince local actors to participate in the process. Across all 10 movements in the analysis, the two most successful frames were those that embodied "economic" arguments and "local environmental" arguments. The former centres around the implications that fossil fuel production has on local fishing/tourism and other jobs, which can be understood as threats to the livelihoods of local community members through reduced employment and food access (socially exclusive) and disruptions in local ecosystems (ecologically exclusive). The "local environmental" frames championed arguments such as protecting local water, soil and air resources from fossil-related pollution, which would also inherently disrupt ecosystems (be ecologically exclusive) and indirectly threaten human livelihoods and health (socially exclusive).

Second, our evidence suggests that successful socio-environmental movements and courts in the Global South have been able to make niche changes in terms of closing down individual coal-fired plants or arresting deforestation.

Third, in order to make a regime change, these movements have often used the courts to challenge state policy and to set a legal precedent. Generally speaking, public protests raise consciousness but may not lead to a change in policy. The court cases are more effective in this regard. The vast majority of court cases led to decisions that could reduce greenhouse gas emissions and had higher "success rates" (rendering verdicts in favour of such mitigative implications in 47/57 cases) than the wider set of actions adopted by the socio-environmental movements. This may suggest that using courts is a more effective way to achieve a better result. However, courts can mostly be used in countries where there is some degree of democracy and space for civic actors, and in countries with some level of institutionalisation. Though these successful court cases in promoting climate mitigation may

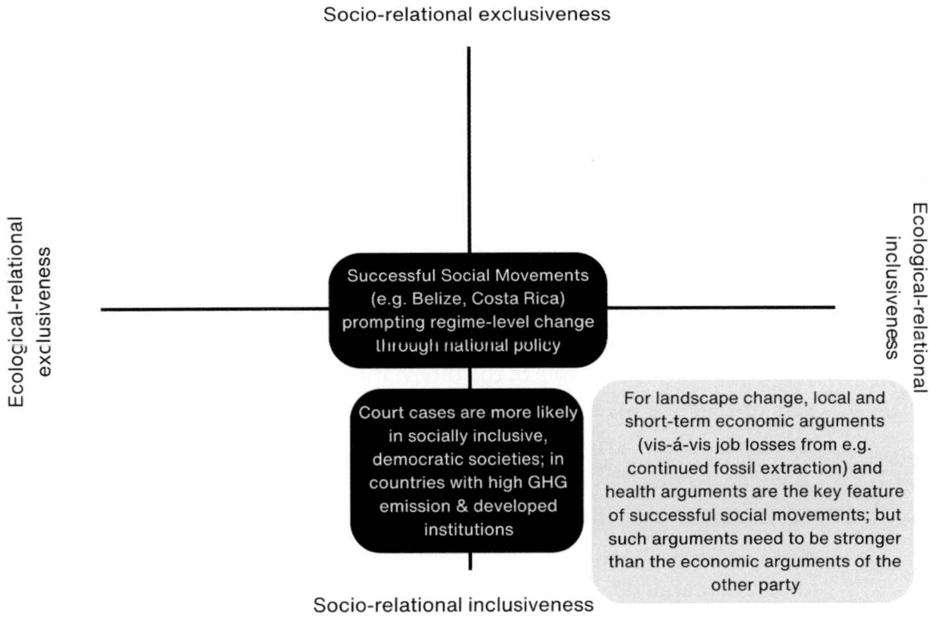

Figure 3.13. Social movements, courts and inclusive transitions.

offer promising prospects as precedents that can eventually shape the regime (and landscape) of fossil fuel-related court cases, it seems that such a regime may be bound by contextual factors, like national economic capacity and national emissions rates. Such court decisions either lead directly to new policy at the state level, or indirectly by encouraging other social actors to build on the legal precedent and to object to fossil fuel production (shaping the regime). Regime change seems to be in its nascent stages, seeing as only 10 environmental movements since 1970 were successful in enforcing policy change, and of those only two did so most impactfully at the national level. Moreover, since none of the six movements from the Global North succeeded in implementing national-level policy change (likely due to the home economies' reliance on fossil fuel exporting and existing infrastructure), continued efforts to catalyse niche-level events (like the two coal-fired power stations closed in Chicago) may have to be both amplified and fine-tuned to eventually rattle the regime level.

Fourth and finally, we are yet to see a landscape change where the rise of social movements and their repertoire of actions, including court cases, eventually influences the relevant landscape (i.e. international/multilateral policies banning coal, oil and gas extraction/production). The role of social movements and courts is key to facilitating the energy transition.

References

Adams, A., & Shriver, T. (2016). Challenging extractive industries: How political context and targets influence tactical choice. *Sociological Perspectives, 59*(4), 892–909. https://doi.org/10.1177/0731121416641683

Amenta, E., Caren, N., Chiarello, E., & Su, Y. (2010). The political consequences of social movements. *Annual Review of Sociology, 36*(1), 287–307. https://doi.org/10.1146/annurev-soc-070308-120029

Andersen, E. A. (2006). *Out of the closets and into the courts: Legal opportunity structure and gay rights litigation.* University of Michigan Press.

Bodansky, D. (2017). The role of the international court of justice in addressing climate change: Some preliminary reflections. *Arizona State Law Journal, 49*, 689–712. https://arizonastatelawjournal.org/wp-content/uploads/2017/09/Bodansky_Pub.pdf

Bond, P. (2018). Social movements for climate justice during the decline of global governance from international NGOs to local communities. In L. Sharachchandra, E. S. Brondizio, J. Byrne, G. M. Mace & J. Martinez-Alier (Eds.), *Rethinking environmentalism: Linking justice, sustainability, and diversity* (pp. 153–182). MIT Press. https://doi.org/10.7551/mitpress/11961.001.0001

Borras, S. M., Jr., & Franco J. C. (2018). The challenge of locating land-based climate change mitigation and adaptation politics within a social justice perspective: Towards an idea of agrarian climate justice. *Third World Quarterly, 39*(7), 1308–1325. https://doi.org/10.1080/01436597.2018.1460592

Bos, K., & Gupta, J. (2016). Inclusive development, oil extraction and climate change: A multilevel analysis of Kenya. *International Journal of Sustainable Development and World Ecology, 23*(6), 482–492. https://doi.org/10.1080/13504509.2016.1162217

Burgen, S. (2014, March 26). Spain's oil deposits and fracking sites trigger energy gold rush. *The Guardian*. https://www.theguardian.com/world/2014/mar/26/spain-oil-deposit-fracking-sites-energy-offshore-gas

Buyse, A. (2018). Squeezing civic space: Restrictions on civil society organizations and the linkages with human rights. *International Journal of Human Rights, 8*, 966–988. https://doi.org/10.1080/13642987.2018.1492916

Carroll, W. K., & Ratner, R. (2010). Social movements and counter-hegemony: Lessons from the field. *New Proposals: Journal of Marxism and Interdisciplinary Inquiry, 4*(1), 7–22.

Centre for Environmental Rights [South Africa]. (2020). Virtual library. https://cer.org.za/virtual-library/judgments

Centre for Food and Adequate Living Rights Limited et al. v Attorney General of the Republic of Uganda et al. (2020). Application no. 29 of 2020. East African Court of Justice.

Chenoweth, E., & Stephan, M. J. (2011). *Why civil resistance works: The strategic logic of nonviolent conflict*. Columbia University Press.

Demaria, F., & Kothari, A. (2017). The post-development dictionary agenda: Paths to the pluriverse. *Third World Quarterly, 38*(12), 2588–2599. https://doi.org/10.1080/01436597.2017.1350821

Earthlife Africa Johannesburg v Minister of Environmental Affairs and Others. (2017). Case no. 65662/16 [2017] 2 All SA 519. South Africa High Court.

ECOLEX. (2020). https://www.ecolex.org/

The Economist. (2019). Democracy Index 2019: A year of democratic setbacks and popular protest. https://www.eiu.com/topic/democracy-index

EIA [Energy Information Administration]. (2019, December 5). US petroleum exports exceed imports in September. https://www.eia.gov/todayinenergy/detail.php?id=42176

EIU [The Economist Intelligence Unit]. (2020). Democracy Index 2019. https://www.eiu.com/public/topical_report.aspx?campaignid=democracyindex2019

EJAtlas [Global Atlas of Environmental Justice]. (N.d.). https://ejatlas.org

Ewbank, L. (2016, June 22). The long road to VRET: Foe's campaign for a Vic renewable energy target. https://yes2renewables.org/2016/06/22/the-long-road-to-vret-foes-campaign-for-a-vic-renewable-energy-target/

Friends of the Earth Melbourne. (2019, June 3). No new fossil fuels in Victoria. https://www.melbournefoe.org.au/no_new_fossil_fuels_vic

Friends of Wild Salmon. (N.d.). LNG development. http://friendsofwildsalmon.ca/campaigns/detail/liquefied_natural_gas_lng_development_85

Ganguly, G., Setzer, J., & Heyvaert, V. (2018). If at first you don't succeed: Suing corporations for climate change. *Oxford Journal of Legal Studies, 38*(4), 841–868. https://doi.org/10.1093/ojls/gqy029

Gbemre v Shell Petroleum Development Company of Nigeria Limited and Others. (2005). Suit no. fhc/b/cs/53/05 [2005] AHRLR 151. Federal Court of Nigeria.

GCA [Global Carbon Atlas]. (2020). CO_2 Emissions, 2018. http://www.globalcarbonatlas.org/en/CO2-emissions

Germanos, A. (2019, April 10). "Monumental": Chicago commits to 100% renewable energy by 2040. *Common Dreams*. https://www.commondreams.org/news/2019/04/10/monumental-chicago-commits-100-renewable-energy-2040

Giugni, M. (1999). Introduction: How social movements matter: Past research, present problems, future developments. In M. Giugni, D. McAdam & C. Tilly (Eds.), *How social movements matter: Past research, present problems, future developments* (pp. 13–33). University of Minnesota Press.

Gomez, J. (2018, January 9). The biggest reef in Americas says "No!" to offshore oil. *Oceana Belize*. https://belize.oceana.org/press-releases/biggest-reef-americas-says-no-offshore-oil/

Grantham Research Institute on Climate Change and the Environment. (2020). Climate Change Laws of the World [Database]. https://climate-laws.org

Gupta, J. (2007). Legal steps outside the climate convention: Litigation as a tool to address climate change. *Review of European Community & International Environmental Law, 16*(1), 76–86. https://doi.org/10.1111/j.1467-9388.2007.00541.x

Gupta, J. (2014). *History of global climate governance.* Cambridge University Press. https://doi.org/10.1017/CBO9781139629072

Gupta, J., & Vegelin, C. (2016). Sustainable Development Goals and inclusive development. *International Environmental Agreements: Politics, Law and Economics, 16*(3), 433–448. https://doi.org/10.1007/s10784-016-9323-z

Harriss-White, B. (2006). Poverty and capitalism. *Economic and Political Weekly, 41*(13), 1241–1246.

Hughes, A. (2018). *Australian Resource Reviews: Black Coal 2017.* Geoscience Australia. https://doi.org/10.11636/9781925297980

Jacobs, R. E. (2005). Treading deep waters: Substantive law issues in Tuvalu's threat to sue the United States in the International Court of Justice. *Pacific Rim Law and Policy Journal, 14*, 103–128.

Klein, N. (2015). *This changes everything: Capitalism vs. the climate.* Simon and Schuster.

Laville, S. (2019, March 22). Top oil firms spending millions lobbying to block climate change policies, says report. *The Guardian.* https://www.theguardian.com/business/2019/mar/22/top-oil-firms-spending-millions-lobbying-to-block-climate-change-policies-says-report

Li, J. (2019). Climate change litigation: A promising pathway to climate justice in China? *Virginia Environmental Law Journal, 37*(2), 132–170.

Lin, A. C. (2014a). Fracking and federalism: A comparative approach to reconciling national and subnational interests in the United States and Spain. *Environmental Law, 44*(4), 1039–1078.

Lin, J. (2012). Climate change and the courts. *Legal Studies, 32*(1), 35–57.

Lin, J. (2014b). Litigating climate change in Asia. *Climate Law, 4*, 140–149.

Markell, D. L., & Ruhl, J. B. (2010). An empirical survey of climate change litigation in the United States. *Environmental Law Reporter, 40*(7), 10644–10655.

Markell, D. L., & Ruhl, J. B. (2012). An empirical assessment of climate change in the courts: New jurisprudence or business as usual. *Florida Law Review, 64*(1), 15–86.

Martine, G., & Alves, J. E. D. (2019). Disarray in global governance and climate chaos. *Revista brasileira de estudos de população, 36*, e0075. https://doi.org/10.20947/S102-3098a0075

Martinez-Alier, J. (2021). Mapping ecological distribution conflicts: The EJAtlas. *The Extractive Industries and Society, 8*(4), 100883.

Martinez-Alier, J., et al. (2014). Between activism and science: Grassroots concepts for sustainability coined by environmental justice organizations. *Journal of Political Ecology, 21*(1), 19–60. https://doi.org/10.2458/v21i1.21124

Mbabazi and Others v The Attorney General et al. (2012). Civil suit no. 283 of 2012. Uganda High Court.

Meijer, B. (2021, January 29). Environmentalists, farmers win Dutch court case over Shell Nigeria spills. *Reuters.* https://www.reuters.com/article/us-shell-nigeria-court-idUSKBN29Y1D2

Mugga, J. I., Gupta, J., & Lefeber, R. (2023). Shaping Africa's climate action through climate litigation: An impact assessment. *Recht in Afrika, 26*(1), 26–54. https://www.nomos-elibrary.de/10.5771/2363-6270-2023-1-26/shaping-africa-s-climate-action-through-climate-litigation-an-impact-assessment-volume-26-2023-issue-1?page=1

Muttitt, G., & Kartha, S. (2020). Equity, climate justice and fossil fuel extraction: Principles for a managed phase out. *Climate Policy, 20*(8), 1024–1042, https://doi.org/10.1080/14693062.2020.1763900

ND-GAIN [Notre Dame Global Adaptation Initiative]. (2019). ND-GAIN Country Index 2018. https://gain.nd.edu/our-work/country-index/

Oilwatch Costa Rica. (2005, August). Ecological debt and oil moratorium in Costa Rica. https://web.archive.org/web/20061007174819/http://www.oilwatch.org/doc/campana/deuda_ecologica/deuda_costarica_ing.pdf

Osa, M. (2001). Mobilizing structures and cycles of protest: Post-Stalinist contention in Poland, 1954–1959. *Mobilization: An International Quarterly, 6*(2), 211–231. https://doi.org/10.17813/maiq.6.2.ml4u77k2370504j0

Osofsky, H. M. (2005). The geography of climate change litigation: Implications for transnational regulatory governance. *Washington University Law Quarterly, 83*, 1789–1855.

Parker, C., & Braithwaite, J. (2003). Regulation. In P. Cane & M. Tushnet (Eds.), *The Oxford handbook of legal studies.* Oxford University Press.

Peel, J., & Lin, J. (2019). Transnational climate litigation: The contribution of the Global South. *American Journal of International Law, 113*(3), 679–726. https://doi.org/10.1017/ajil.2019.48

Peel, J., & Osofsky, H. M. (2013). Climate change litigation's regulatory pathways: A comparative analysis of the United States and Australia. *Law & Policy, 35*(3), 150–183.

Peel, J., & Osofsky, H. M. (2015). *Climate change litigation: Regulatory pathways to cleaner energy.* Cambridge University Press.

Pérez-Rincón, M., Vargas-Morales, J., & Martinez-Alier, J. (2019). Mapping and analyzing ecological distribution conflicts in Andean countries. *Ecological Economics, 157*, 80–91.

Piggot, G. (2018). The influence of social movements on policies that constrain fossil fuel supply. *Climate Policy, 18*(7), 942–954. https://doi.org/10.1080/14693062.2017.1394255

Poppe, A. E., & Wolff, J. (2017). The contested spaces of civil society in a plural world: Norm contestation in the debate about restrictions on international civil society support. *Contemporary Politics, 4*, 469–488. https://doi.org/10.1080/13569775.2017.1343219

Presthus, R. Y. (1974). *Elites in the policy process*. Cambridge University Press.

Preston, B. J. (2016). The contribution of the courts in tackling climate change. *Journal of Environmental Law, 28*(1), 11–17.

Quist, L. M. (2019). Fishers' knowledge and scientific indeterminacy: Contested oil impacts in Mexico's sacrifice zone. *Maritime Studies, 18*(1), 65–76.

R (on the application of Friends of the Earth Ltd and others) v. Secretary of State for Business Energy and Industrial Strategy. (2022). EWHC 1841 (Admin). High Court of Justice of England and Wales.

Rajamani, L. (2016). Ambition and differentiation in the 2015 Paris Agreement: Interpretative possibilities and underlying politics. *International & Comparative Law Quarterly, 65*(2), 493–514. https://doi.org/10.1017/S0020589316000130

Rico. (2019, 28 February). Costa Rica extends moratorium on oil exploration. *Q Costa Rica*. https://qcostarica.com/costa-rica-extends-moratorium-on-oil-exploration/

Ritchie, H. (2017). Fossil fuels. Our World in Data. https://ourworldindata.org/fossil-fuels

Sabin Center for Climate Change Law. (2020). Global Climate Change Litigation [Database]. https://climatecasechart.com/non-us-%20climate-change-litigation/

Sands, P. (2016). Climate change and the rule of law: Adjudicating the future in international law. *Journal of Environmental Law, 28*(1), 19–35.

Savaresi, A., & Auz, J. (2019). Climate change litigation and human rights: Pushing the boundaries. *Climate Law, 3*, 244–262. https://doi.org/10.1163/18786561-00903006

Save Lamu and Others v National Environmental Management Authority and Amu Power Co. Ltd. (2019). Tribunal appeal no. NET 196 of 2016 [2019] eKLR. National Environmental Tribunal Kenya.

Scheidel, A., et al. (2020). Environmental conflicts and defenders: A global overview. *Global Environmental Change, 63*, 102104.

Scheidel, A., & Schaffartzik, A. (2019). A socio-metabolic perspective on environmental justice and degrowth movements. *Ecological Economics, 161*, 330–333. https://doi.org/10.1016/j.ecolecon.2019.02.023

Setzer, J., & Benjamin, L. (2020). Climate litigation in the Global South: Constraints and innovations. *Transnational Environmental Law, 9*(1), 77–101. https://doi.org/10.1017/s2047102519000268

Setzer, J., & Byrnes, R. (2019). Global trends in climate change litigation: 2019 snapshot. London School of Economics and Political Science.

Setzer, J., & Vanhala, L. C. (2019). Climate change litigation: A review of research on courts and litigants in climate governance. *WIREs Climate Change, 10*(3), e580. https://doi.org/10.1002/wcc.580

Skrentny, J. D. (2006). Policy-elite perceptions and social movement success: Understanding variations in group inclusion in affirmative action. *American Journal of Sociology, 111*(6), 1762–1815. https://doi.org/10.1086/499910

Snow, D. A. (2013). Framing and social movements. In *The Wiley-Blackwell encyclopedia of social and political movements*. Wiley. https://doi.org/10.1002/9781405198431

Soto, L., Alejandro, E., Herrera, R., Montoya-Marquez, J., Ruiz, R., Corona, A., & Monterroso, C. (2009). Biodiversidad Marina en la Sonda de Campeche. In L. A. Soto & M. del Carmen González-García (Eds.), *Pemex y la salud ambiental de la Sonda de Campeche, Mexico* (pp. 265–300). Battelle Memorial Institute, Instituto Mexicano del Petróleo, Universidad Autónoma Metropolitana and Universidad Nacional Autónoma de México.

Spanish News Today. (2020, April 14). Spanish government plans to ban fracking. https://spanishnewstoday.com/spanish-government-plans-to-ban-fracking_1396024-a.html

Temper, L., Del Bene, D., & Martinez-Alier, J. (2015). Mapping the frontiers and front lines of global environmental justice: the EJAtlas. *Journal of Political Ecology, 22*(1), 255–278. https://doi.org/10.2458/v22i1.21108

Temper, L., et al. (2018). The global environmental justice atlas (EJAtlas): Ecological distribution conflicts as forces for sustainability. *Sustainability Science, 13*(3), 573–584. https://link.springer.com/article/10.1007/s11625-018-0563-4

Temper, L., et al. (2020). Movements shaping climate futures: A systematic mapping of protests against fossil fuel and low-carbon energy projects. *Environmental Research Letters, 15*(12), 123004. https://doi.org/10.1088/1748-9326/abc197

Temper, L., Yánez, I., Sharife, K., Godwin, O., & Martinez-Alier, J. (Eds.). (2013, May). *Towards a post-oil civilization: Yasunization and other initiatives to leave fossil fuels in the soil.* EJOLT report no. 6. http://www.ejolt.org/wordpress/wp-content/uploads/2013/05/130520_EJOLT6_Low2.pdf

Tramel, S. (2016). The road through Paris: Climate change, carbon, and the political dynamics of convergence. *Globalizations, 13*(6), 960–969.

US Department of Energy. (2019). Frequently asked questions (FAQs): US Energy Information Administration (EIA). https://www.eia.gov/tools/faqs/faq.php?id=104&t=3%0Ahttp://www.eia.gov/tools/faqs/

Vanhala, L. (2012). Legal opportunity structures and the paradox of legal mobilization by the environmental movement in the UK. *Law & Society Review, 46*(3), 523–556. https://doi.org/10.1111/j.1540-5893.2012.00505.x

Walker, C. (2016). The campaign against new coal and gas in Victoria: The story so far. Friends of the Earth Melbourne. http://www.melbournefoe.org.au/coal_and_gas?utm_campaign=gasbanwin2017&utm_medium=email&utm_source=friendsofearthmelbourne

Walker, E. T., Martin, A. W., & McCarthy, J. D. (2008). Confronting the state, the corporation, and the academy: The influence of institutional targets on social movement repertoires. *American Journal of Sociology, 114*(1), 35–76. https://doi.org/10.1086/588737

Wedy, G. (2017). *Climate legislation and litigation in Brazil*. Sabin Centre for Climate Change Law, Columbia Law School. https://ssrn.com/abstract=3052226

Wilensky, M. (2015, February). *Climate change in the courts: An assessment of non-US climate litigation*. Sabin Centre for Climate Change Law, Columbia Law School. https://doi.org/10.7916/D80P0Z66

Wilson, B. M., & Cordero, J. C. R. (2006). Legal opportunity structures and social movements: The effects of institutional change on Costa Rican politics. *Comparative Political Studies, 39*(3), 325–331.

World Bank. (2020). GDP, PPP (current international $), Data, 2019. https://data.worldbank.org/indicator/-NY.GDP.MKTP.PP.CD

WWF [World Wildlife Fund]. (2018). Gran alianza contra el gas en España. https://www.wwf.es/?uNewsID=47440

4. Latin America, Ecuador and Social Mobilisation for LFFU

Carolina Valladares and Barbara Hogenboom

Abstract

Latin American social movements have globally been at the forefront of resisting fossil fuel projects and developing arguments and approaches to leave fossil fuels underground. This chapter analyses the LFFU arguments and approaches of these bottom-up actors in Ecuador and other Latin American countries. From the perspective of the politics of ideas, the chapter looks into the ways in which bottom-up actors operating in countries with major development challenges envision LFFU and an energy transition, and how they deal with dominant imaginaries of fossil fuels as a source of development. Since the turn of the century, social mobilization for LFFU in Latin America has strengthened and connected to various struggles for social and environmental justice. Agendas of various civil society actors come together: environmental protection, territorial rights, women's rights, protecting land and water for agriculture, and countering unsafe living and working conditions. The analysis shows that Latin American LFFU proposals are linked to broader processes for change, including alternative views on development and rights.

Keywords: fossil fuels, resistance, social movements, politics of ideas, Latin America

4.1. Introduction

Latin American social movements have been at the forefront of resisting fossil fuel projects and developing arguments and approaches to leave fossil fuels underground. Next to numerous local contestations, national and

Gupta, J., Hogenboom, B., Rempel, A. & Olofsson, M. (eds), *Leaving Fossil Fuels Underground: Actors, Arguments and Approaches in the Global South and Global North.*
Taylor & Francis Group, 2024
doi: 10.5117/9789048560370_CH04

regional networks have emerged since the 1980s, through which proposals to leave fossil fuels underground are exchanged. These actors are part of the wider regional movement for socio-environmental justice and alternative(s to) development that influences policy and legislation in Latin American countries and inspires debates and social movements on LFFU and development in the Global South and the Global North. In this chapter we focus on the ways in which social movements are driving LFFU initiatives in Latin America. In particular, we will examine the arguments and approaches of these bottom-up actors, and the ways in which they engage various actors to realise LFFU. We take a key interest here in the politics of ideas, and the question of how bottom-up actors operating in countries with major development challenges envision LFFU and an energy transition. There are important lessons to learn from the ways in which they have not only mobilised multiple agendas against fossil fuels, but also state and other actors, and alternative views on development and rights. This is key to our understanding of their role in the energy transition.

While the availability of oil, coal, gas and shale gas reserves varies from country to country, overall Latin America has massive reserves (see 1.6.2) and it is the world's second-largest oil-producing region, after the Middle East (Valdivia & Lyall, 2018). Extractive operations increasingly expand and are moving into more remote areas such as the Amazon (Bebbington et al. 2018b), but also into the (deep) sea in countries like Brazil, Guyana and Suriname, and into new technologies such as fracking, especially in Argentina. Latin American protests against fossil fuel projects were initially focused on the ecological damage and negative territorial consequences of specific local projects, but increasingly this criticism fed into broader debates on the downsides of development based on extractivist sectors and models. With the increasing scientific proof of the urgency to halt climate change, another key argument was added to the already existing LFFU approaches. The case of Ecuador is an important example of the rise of LFFU actors, arguments and approaches, indicating the potential and controversiality of LFFU in a country with large fossil fuel reserves. Since the start of the exploitation of oil in the early 1970s, this resource has been central to its economy and overall development model, and resource nationalism has been a constant factor, even at times of neoliberal restructuring. As its large oil fields are in the Amazon, for 50 years the projects have been a driving force behind rapidly opening the rainforest and bringing in mega infrastructure to connect subsoil resources to global markets (cf. Larrea, 2022).

This chapter on Latin American bottom-up LFFU initiatives, with special attention on the case of Ecuador, builds on a literature review and extensive research in and on Ecuador, and is structured as follows. It starts with a presentation of the analytical framework we apply to assess bottom-up actors, arguments and approaches, building on key elements of Chapter 2 (see 4.2). Next, we briefly review the social and political imaginaries of the fossil fuel sector bringing development and justice to the region (see 4.3). Then we examine key LFFU proposals and broader agendas for change, including the mobilisation of multiple agendas, different actors and alternative views on development and rights (see 4.4). In the concluding section, we discuss which LFFU initiatives have the most potential in and beyond the region.

4.2. Building on the analytical framework

With our focus on the role of Latin American social movements as key actors in the promotion of LFFU ideas, we are especially interested in promising approaches and arguments that reach beyond the local sites of extraction and resistance, and which aim at more comprehensive agendas for socio-environmental justice, inclusiveness and sustainability. This interest connects closely to the notion of the politics of ideas that was introduced in Chapter 2. As mentioned there, this notion invites us to look at natural resource politics beyond merely economic interests and power relations, and to consider the influence of "the ideas about that resource" (Bebbington et al., 2018a, p. 204). For social movements, developing persuasive new ideas and proposals is an indispensable political task. As we will see in Section 4.4, making the local effects of the fossil fuel industry and the local struggles in marginalised areas visible in society at large is an important first step in the process of developing comprehensive arguments and approaches that not only resist certain projects but also propose and up-scale alternatives to fossil fuel extraction. As a second step—in line with Bebbington's insightful white paper "Natural Resource Extraction and the Possibilities of Inclusive Development: Politics across Space and Time" (2013)—building coalitions between (various) NGOs, social groups and movements, scholars and other actors allows these proposals to be brought forward in public opinion and policymaking. On several occasions of bottom-up LFFU initiatives in Latin America, this has indeed proven to be the case. Such a comprehensive approach to the development of coalitions and proposals, which in some cases also includes (the views of) Indigenous groups, ecologists, oil sector workers and feminist organisations, results in a combination of various struggles for justice as well as a broad understanding of inclusive development. In the case of Latin American LFFU struggles, we also often see a third step of connecting these new ideas, approaches, arguments and coalitions to global climate debates and action and funding schemes, which relates to the relational dimension of inclusive development that is discussed in Chapter 2 (see 2.2).

For our analysis it is important to recognise that bottom-up mobilising (for) LFFU ideas in Latin America, including local activism, coalition-building and court cases (see Chapter 3), tends to be very challenging. Evidently there are enormous public and private, national and international interests at stake in the region's fossil fuel sector, especially in countries like Brazil, Mexico, Venezuela, Colombia, Argentina and Ecuador. An international political economy assessment of this sector is beyond the scope of this chapter, but it does need mentioning that next to the historically grown investments and loans from US, Canadian and European entities, since around 2005, Chinese capital has also come in, adding another layer of powerful interests and influence (cf. Hogenboom, 2017). Next, the region has a notoriously bad track record of violations of the human rights of environmental defenders, as repeatedly shown in reports of international human rights NGOs. According to the Inter-American Commission on Human Rights and the Latin American country branches of the Office of the United Nations High Commissioner for Human Rights, Latin America is the most dangerous region for environmental defenders, "particularly women and peasant, Indigenous, and Afro-descendant leaders and their communities ... [when they] express their concern about the impact of certain business activities on land and territory" (IACHR et al., 2022). These local spheres and acts, threats and oppression are in many cases accompanied by a national dominant framework in which communities and NGOs resisting fossil fuel projects are easily perceived as egoistic—prioritising their own interests over that of the nation—and anti-developmental—prioritising nature or the environment over social and economic needs of society. As a result, LFFU actors are involved in a constant struggle to counter (simplistic) accusations and not be pushed away from the debate on what is sustainable development and social and environmental justice. However, as we will see, participating in this debate is crucial for these social movements. The bottom-up LFFU movement is involved in what we can call "beyond-resistance strategies"—a clear case of social movements being crucial agents in imagining the "not yet" (Dinerstein, 2015). To understand this dimension of Latin American LFFU movements, this chapter also addresses the role of imagination and cultural politics.

4.3. A century of fossil fuels, development and justice debates

From the start of their industrial extraction, fossil fuel operations deeply transformed nature, society and the political economy. One of the earliest experiences was in Mexico, in the early 20th century, where oil production gave way to a major transformation in land use and social organisation, which brought about environmental destruction, marginalisation of Indigenous groups as well as tense labour relations (Santiago, 2006). Even though, over time and space, this "story" has been repeated multiple times, the dominant ideas about fossil fuels in society and politics link these resources with progress and notions of inclusive development. Resource nationalism has been—and arguably still is—the most powerful idea in Latin America's mineral governance. "Resource nationalism is the notion that natural resources should be managed above all for the needs of 'the nation,' and therefore should be controlled by the state for 'the people'" (Bebbington et al., 2018a, pp. 207–208). Popular (and) political expressions of this idea have been rather persistent over time, and it comes as no surprise that in the course of the 20th century, Latin American oil countries nationalised this sector. In Venezuela, the idea of oil bringing development was already widespread in the 1930s with popular concepts such as "sowing oil" (*sembrar el petróleo*—an interesting but also puzzling agricultural metaphor, considering the polluting nature of oil), being promoted in different wording during the 2000s and the "oil boom" under President Hugo Chavez. In several countries, historical murals in national government buildings depict minerals as the source of the national progress that need to be protected against greedy foreign powers (cf. Pellegrini, 2018, on Bolivia). Popular slogans during mid-20th-century mass meetings about "we, the people" as owners of oil and gas (e.g. Brazil's *o petróleo é nosso*; and *el gas para los bolivianos*), reappeared in the early 21st century, and this call to undo neoliberal reforms was to some extent successful. Also, the old image of exultant presidents dipping their hands in oil became, again, part of the political imaginary among progressive presidents of the 21st century, such Venezuela's Chavez and Brazil's Luiz Lula da Silva.

While fed by politicians, oil and other fossil fuel entrepreneurs and state companies, and the sector's workers' unions, this imagination of oil and gas as sources of national wealth, development and social justice has become deeply entrenched in society. For instance, in the case of Venezuela, oil is engrained in collective imaginaries and sociocultural patterns. The country is often seen as a textbook example of the resource curse, with a boom and bust economy and many institutional and social problems.

Nevertheless, most citizens in Venezuela's highly unequal society feel an affective relationship to oil and assume that national development has to be oil-based. Oil revenues allowed for massive imports and low taxes, for social programmes and subsidised food and fuel. As a result, next to a rentier state, Venezuela has also become a rentier society (Peters, 2017, p. 53). For poor people living in the capital city of Caracas, far away from the oil fields, oil has sociocultural properties related to citizenship, nationhood and justice (Åsedotter Strønen, 2017, p. 323). Also among the middle and upper classes, oil has shaped collective thinking in such a way that hardly anyone can imagine a national development model that is not based on oil (Peters, 2017, p. 62).

Critical scholars and activists have denounced the revival of traditional ideas and imaginaries regarding fossil fuels and other minerals as neo-developmentalism and neo-extractivism. They have coined this tendency among (progressive and neoliberal) Latin American governments as the Commodity Consensus (Svampa, 2015) and the extractive imperative (Arsel et al., 2016). In particular, the lack of governmental attention for the effects of mineral extraction on nature and on local groups was hardly foreseen. "Si eres tan progresista ¿Por qué destruyes la naturaleza?" ("If you are so progressive, why do you destroy nature?"), was the title of an article by Eduardo Gudynas (2010), one of Latin America's leading environmental experts. Politicians who originally claimed to respect the rights of Indigenous peoples and peasants, and who embraced enactments summoning Pachamama (Mother Earth) and the rights of nature, later on, allowed mineral extraction to rapidly expand and even suppressed criticism and protest by affected groups.

For decades, social movements in Latin America have been putting the spotlight on the region's energy model and the role of fossil fuels in the development model. Networks of social organisations, communities, labour unions, Indigenous movements and scholars have been asking the question, For whom and what is the energy for? As shown in 1.6.2, most of Latin America's extracted oil, coal and gas are for export. Despite substantial revenues there is still a widespread social and economic inequality as well as inequality of access to energy. A large share of the extracted energy and related revenues does not necessarily improve the living conditions of impoverished populations or help to build a sustainable and inclusive economy, but the externalities of those exports, such as deforestation, pollution, water scarcity, health problems and sometimes violence, immediately affect the population around and beyond extraction sites (Cardoso, 2015) and impact future generations.

In theory, according to most Latin American constitutions, it is "the people" who own all subsoil reserves, which in practice means that the Latin American state controls all fossil fuel resources, supposedly to serve the interests of the nation. However, under such public ownership, the excessive (potential) earnings from mineral wealth often corrupt individuals, institutions and systems. This is also partly due to the worldwide problem of secrecy in the sector, because, as Ross (2012, p. 244) stresses: "Most of the oil world is hidden for public view," which is a key driver of the oil curse. In Latin America, despite some variations, corruption is a generalised phenomenon around the extractive sector: it has a long history and happens at all levels, even in countries that score high in transparency rankings (Gudynas, 2019). The region-wide Odebrecht scandal around infrastructural projects showed that corruption is not limited to the extractive sector, but among the long list of recent corruption cases, many are related to mining and especially oil. Interestingly, in this context of widespread corruption in the fossil fuel system, political-economic elites, such as in the case of Ecuador, legitimise their claims over oil rights and the petro-state apparatus in moral terms, "rooted in a widely-shared moral economy of oil" (Lyall, 2018, p. 394).

4.4. Latin American LFFU proposals and broader agendas for change

4.4.1. Multiple civil society actors

In Latin America, the calls of social movements to LFFU are connected to various struggles for social and environmental justice. Next to environmental and ecologist groups, Indigenous movements and organisations of feminists and peasants also address fossil fuel and energy issues, and more broadly the extractivist development model. Locally, in fossil fuel extraction sites, agendas of various civil society actors come together: environmental protection, territorial rights, land and water for agriculture, and countering unsafe and exploitative living and working conditions, especially for women. Throughout the region, national environmental organisations support the often-dispersed local struggles against extractivist (mega-)projects (Silva et al., 2018) and aim to change national policies that cause or neglect the damage of fossil fuel projects. As Svampa (2015) points out, together with intellectuals and experts, Indigenous and peasant movements and cultural collectives have recently formed a complex social fabric. Part of this trend is "the environmentalisation of Indigenous and campesino struggles and the emergence of a Latin American environmental thinking" (Leff in Svampa, 2015, p. 68). It is important to stress, however, that Latin American environmentalism has a long history and strong regional tradition, dating back to the 19th century. From the 1970s onwards, scientific, governmental and social environmentalism strengthened, and especially around the Rio 1992 conference, the region's popular environmentalism and environmental movement became visible and vocal (Martinez-Alier et al., 2016).

The Indigenous movement, which has arguably been the region's most important source of social mobilisation since the 1990s (cf. Yashar, 2005), has played a central role in various LFFU cases. What stands out in their approach is that next to their general claims for recognition, political participation and an end to systematic institutional discrimination, they expose and use the profound territorial and cultural dimensions of fossil fuel extraction to halt planned projects. In Ecuador, this is, for example, clear in the mobilisation of Kichwa Indigenous people in Sarayaku after they found out that the state had given part of their territory in concession to an oil company without any consultation. They notified the government of their opposition to oil operations on their ancestral lands, organised to avoid the company entering the territory and started a court case. In 2012, the Inter-American Court of Human Rights (IACHR) ruled in their favour and pointed out that all signatories of the American Convention on Human Rights are bound to seek the free, prior and informed consent of Indigenous peoples prior to the start of activities that may affect their territories and lives (CEJIL, 2012). This important decision was sustained on arguments such as Indigenous people's tradition of collective landownership, their territory as a guarantee for their survival and their right to cultural identity (IACHR, 2010). The cultural and spiritual significance of territory in Indigenous struggles for LFFU was also articulated strongly in the U'was resistance against an oil project in the mid-1990s in Colombia. "Oil is Earth's blood and extracting it would dry out Mother Earth," they declared as they threatened with collective suicide (Serje, 2003, our translation). Through regional networks of Indigenous movements such local strategies, arguments and experiences are exchanged. Representatives of Ecuadorian Indigenous people affected by the ChevronTexaco oil spills in the Amazon, for instance, visited Mapuche people fighting fracking projects of the same company in Argentina. Also, Indigenous initiatives take shape across national borders, such as the "Sacred Watersheds" programme in the Peruvian and Ecuadorian Amazon, through which extractive activities are jointly monitored and exposed.

In recent years, women have taken a more prominent role in LFFU struggles, especially impoverished and Indigenous women from towns or territories where social and economic dynamics revolve around the oil industry, which hurts their health, safety and autonomy (Caretta & Zaragocin, 2020; Cielo et al., 2016). In Ecuador, a coalition of Amazon Indigenous women from seven nationalities opposes the expansion of extractive frontiers in their territories, including new oil concessions. They have spearheaded marches from their hometowns to the capital, occupied public buildings, talked with authorities and delivered manifestos demanding their rights to self-determination. One of the reasons these "Amazon Women," as they call their coalition, are taking the lead in LFFU, they explain, is because men often negotiate agreements with (oil) companies without consulting women (El Universo, 2013). Their manifesto also points to gender-based violence in extractive areas, and states: "[We Indigenous women] are 50% of the Indigenous population, carriers of life, who take care of families and Mother Earth" (Amazon Watch, 2018). This exemplifies the regional rise of Indigenous feminism, in which actions for women's rights and territorial rights are interlaced with Indigenous epistemologies and cosmovisions (Hernández Castillo, 2010). More broadly also, ecofeminism, which encompasses critiques of the economic system and the simultaneous exploitation of female bodies and nature, is influencing regional debates on the need for a different development and energy model (Svampa, 2015).

It is at the intersections of gender, ethnicity and class as well as rural–urban and local–global agendas that relevant new LFFU approaches and arguments are emerging in Latin America. Alliances with urban ecologists and feminist organisations, scholars and INGOs have strengthened the local women's struggles (Moreano, 2014), and have helped to raise national and international visibility of women in the Amazon mobilising for LFFU. The increasing international attention for female Indigenous leaders from the Amazon shows in prominent news outlets. *Time* magazine placed Waorani leader Nemonte Nenquimo among the hundred most influential people of 2020 (Time, 2020). She had a leading role in the landmark victory of Waorani communities against the Ecuadorian government. The highest court ratified a legal sentence that bans oil plans in 180,000 hectares of forest of their ancestral territory due to the government's breach of free, prior informed consent (see Figure 4.1). Similarly, *Vogue* counts Sarayaku environmental activist Nina Gualinga among the "7 inspiring young activists working to save the planet" (Vogue, 2021). These and other Amazon women see their local struggles firmly connected to the global climate crisis. As one of them stated during a protest in Ecuador: "The jungle is not just for Waorani, but for the world. We want to leave this tree and river for the future generations" (2018, authors' fieldwork, own translation).

Figure 4.1. Waorani people celebrating their historic legal victory protecting a half-million acres of their rainforest territory from oil exploitation in Puyo, Pastaza, Ecuadorian Amazon.
Photo: Mitch Anderson/Amazon Frontlines.

4.4.2. Mobilising state and non-state actors for LFFU

While in some instances a local group or social movement has managed to halt a planned fossil fuel project by themselves, through protest and legal means, usually to successfully advance an LFFU agenda requires multi-actor coordination. In particular, when it comes to developing a pre-emptive strike or proactive strategy, state and other actors also need to be mobilised since this implies changing policies and regulations. At the local level, Argentina has seen a wave of municipal bans on fracking, following concerns about the expansion of this technology to extract shale gas. When in 2010 large shale reserves were discovered in northern Patagonia (Vaca Muerta), NGOs like OPSUR warned about the environmental risks. In areas where next to activists also citizens, experts, entrepreneurs and politicians became concerned, proactive decision-making moved ahead rather fast. In 2013, social organisations and local public officials of the municipality of Cinco Saltos, in the province Río Negro, succeeded in realising the first ban on fracking in Latin America. Since then, around 50 municipalities have banned fracking (EJES, 2018; OPSUR, 2016). Especially in areas with strong agricultural or tourist sectors, moratoria were adopted as local politicians expected fracking would generate substantial social conflict and limited economic gains (Christel & Mariano, 2018). Also, in Colombia the plans to start fracking projects have brought together various groups—environmentalists, oil workers, artists, scholars, etc.—into the Alliance for Fracking-free Colombia that has effectively influenced public opinion against the idea of fracking in the country since 2013 (Roa, in Jiménez, 2020).

Ecuador's history of social resistance and alternative policy proposals for the oil industry has stood at the cradle of many LFFU ideas in the region and beyond. Among the experiences, the lawsuit of Amazon Indigenous persons and peasants against ChevronTexaco and the socio-environmental impact of its operations stands out (Kimerling, 2006, see also 1.6.2). A few years after the lawsuit was filed, the transnational network Oilwatch, founded in 1996 by the Ecuadorian organisation Acción Ecológica and Earth Rights Action from Nigeria, for the first time proposed an oil moratorium. The observatory and resistance network Oilwatch is present in oil producer countries in the Global South. When Oilwatch's declaration was presented at COP3 at Kyoto in 1997, it was seen as a radical proposal (see Chapter 2). Around a decade later, social organisations and key actors shaped it into the Yasuní–ITT proposal to LFFU launched by the Ecuadorian government to the world; the first of its kind. The initiative was nurtured from years of social struggles of those affected by oil extraction in the northern Amazon and by the first coalitions with ecologist NGOs like Acción Ecológica and international networks. Key actors with a background of years of struggles as activists, intellectuals and scholars acted as brokers to connect with policymakers at the beginning of the new government initially supported by social movements. The initiative mobilised the message that LFFU is the only way to address climate change effectively (Acosta, 2013; Larrea & Warnars, 2009). Even after the failure of Ecuador's proposal to the world (2007–2013)—to keep the oil reserves under a part of Yasuní National Park (YNP) in exchange for the industrialised countries compensating 50% of the potential revenues—it is an unprecedented and eloquent example of how societies are pushing governments and international institutions to think out of the box to make LFFU happen (see 8.3.2 for details of the Yasuní–ITT proposal and its institutional and funding mechanisms).

To keep the oil underground in Yasuní became the symbol of new efforts and ideas to move to a "post oil" civilization (Temper et al., 2013). *Yasunizar*—Yasuní as a verb—became a symbol, a utopia for an energy transition which necessarily implies a civilizational transformation. An energy transition is not merely swapping fossil fuels for renewable sources while continuing with the extensive and unequal use of energy. Our food and transport systems, the way cities are built, are all based on fossil fuel energy. A shift implies rethinking the way we understand, produce, distribute and use energy. Furthermore, it also implies an awareness of the way the energy model is based on the exploitation of nature and women; the subordination of reproduction and care labour to production. From the disappointment over the government's decision to call off the initiative emerged the collective Yasunidos. The marches and protests in some of Ecuador's main cities condemning the cancellation of the Yasuní–ITT Initiative exposed something so far unseen: the extent to which the idea to LFFU had become internalised in new generations and a broad group of citizens. This group formed mostly, but not exclusively, of young people in urban areas, organised to challenge via a referendum the decision to exploit the ITT oil block in the Yasuní. Yasunidos became a space where a broad coalition of people came together to contribute from different knowledge, geographies and strategies to build a colourful, creative and effective mobilisation to LFFU. Rural and urban activists, artists, students, scholars, social organisations and citizens, many of which had never before been mobilised, gathered on the streets and collected signatures to support the referendum to LFFU. Volunteers around the country collected 600,000 signatures, which is the equivalent of 5% of the voting population (see Figure 4.2). The referendum request, however, was rejected by the national government, which was a polemic decision that years later public institutions recognised as a fraud (TCE, 2022; Castro, 2021). Nevertheless, Yasunidos was the beginning of a new style of mobilising environmental debates in Ecuador, with a strong focus on LFFU. New languages and interests of animalists, vegans, activist cyclists, artists, to mention a few, met together with social actors like Indigenous groups, ecologists, feminists, scholars, to create alliances, public appearances, and communication strategies. Despite the social mobilisation, the consecutive governments have entered deeper into the YNP to extract oil. Broad coalitions of actors continue to contest oil extraction in YNP and strive for the "Yasunization" of YNP and other territories threatened by extractive projects.

The struggles of LFFU, by different actors in Ecuador, have achieved important victories and imply a structural questioning in how we understand "development." In the last years, social movements have successfully halted the government's determination to expand oil extraction. The abovementioned Waorani victory in 2019 that prevented oil extraction in their ancestral lands sets important jurisprudence on extractive industries and Indigenous rights. In line with *Yasunizar* (to Yasunize) other places, in Azuay province, a broad mobilisation achieved two popular referenda where people overwhelmingly voted against large-scale mining in fragile *páramo* ecosystems. In 2022, judgements of the Constitutional Court and

Figure 4.2. Yasunidos deliver the signatures required to request a referendum to LFFU in Yasuní National Park to the National Electoral Council.
Photo: Yasunidos.

the Electoral Tribunal made it possible again to take up the referendum rejected to Yasunidos in 2013 to LFFU in YNP. The same court declared unconstitutional the expansion of oil activities in the off-limits zone in the Amazon where uncontacted Indigenous groups live, which overlaps with one of the country's largest oil reserves. Moreover, the government and the Indigenous movement agreed to a temporary moratorium to the oil bid that seeks to expand oil activities to central and southern Amazon. This oil moratorium will last until a law for free prior informed consent has been approved (Serrano, 2022). Importantly, public opinion has become more receptive to environmental issues over the years, thanks to social actors and coalitions that have been keen to mobilise debates around extractivism, development and LFFU. Their ideas, claims and proposals have helped to shift public opinion and, ultimately, policies that match with the times and concerns of the public.

As will be discussed in more detail in Chapter 7, in Latin America also, several other initiatives for fossil fuel moratoria and bans have emerged, which involve interesting "supply-side" policies for LFFU and mitigating climate change (Tudela, 2020). National oil moratoria were adopted in Costa Rica and Belize to prevent exploration and production along the coast, while in Mexico oil bans apply for several protected areas (biosphere reserves). Environmental movements and conservation groups are not alone. In the case of Belize, we see that the idea of a moratorium became influential when it was bundled with a multi-actor coalition (see Chapter 3). The tourist industry was an important actor supporting a ban on oil extraction out of concerns about the effects of an oil spill. In Costa Rica, the government was a more prominent actor, with concerns for how to protect the ecotourism sector and green image of the country. In both countries, persuasive elements include the notion of (potential) *economic costs* of fossil fuel extraction (see Chapter 3). On other occasions, civil society actors aim to mobilise this notion by putting pressure on financial institutions to stop funding fossil fuels. In Ecuador, environmental NGOs with their local and international contacts and coalitions put pressure on European banks to withdraw from financing fossil fuel projects in Ecuador (Reuters, 2021). (Similar developments can be found in recent ideas and policies on mining projects and mining moratoria.)

4.4.3. Mobilising alternative views on development and rights

If extracting fossil fuel is not an option, the exit is to transform the way we understand development. That is what social movements and organisations in Latin America propose. The long-standing criticism of Latin American social movements to fossil fuel extraction has evolved with the years to a comprehensive objection of the current development model. At the core of this model is the permanent and ever-increasing consumption of fossil energy to keep economic growth. When oil, gas and coal projects started decades ago, the first concerns had to do with environmental damage. Shortly after, in the light of the neoliberal reforms, there was growing resistance to the privatisation of fossil fuel reserves. Then, Indigenous groups made the territorial and identarian dimensions of the conflicts for fossil fuels visible. In recent years, social movements on the continent have woven regional and international networks to discuss and build up alternatives to the social environmental challenges of the region, also in the face of the climate crisis. In the Grupo Permanente de Trabajo de Alternativas al Desarrollo, for instance, scholars and activists from the continent hold periodic debates, organise events and publish their reflections on the development model and alternatives (Lang & Mokrani, 2011). There is a permanent exchange and discussions among Latin American and international organisations, social movements, activists and engaged scholars committed to build a "post-development" agenda (Kothari et al., 2019; Escobar, 2010). Oilwatch Latin America has been one of the spaces where environmental organisations from different countries periodically interact in forums but also in in situ visits. In these exchanges, the calls to

LFFU necessarily imply a call for overcoming the concept of development. The Second Latin American Exchange of Fishermen and Fisherwomen against the Exploitation of the Sea (2° Intercâmbio Latino Americano de Pescadores e Pescadoras contra a Exploração do Mar) took place in 2019 in Brazil as part of the "Nem um poço a mais, por áreas livres de petróleo" ("No more pits, for oil-free areas") campaign. There were participants from eight regions of the country and Latin American activists from seven countries. More recently, social movements regional networks launched the "Latin American and Caribbean Platform for Climate Justice." The participants discuss solutions to the climate crisis from the bottom up, as opposed to those of the international climate negotiations. They call for LFFU and to build other society models (Plataforma Latinoamericana, 2021). Initiatives like "The Amazon We Want" seek to weave together scientific, Indigenous and non-Indigenous knowledges. This science panel for the Amazon—"the first high-level science initiative dedicated to the Amazon"—aims to advance arguments for the conservation of this rainforest and for tackling the destructive trends that threaten it (Science Panel for the Amazon, 2021). Latin American social movements strive to build proposals and alternatives from the experiences, knowledge and needs of the people on the continent, adjusted to the diversity of cultures and interactions with the environment.

Some groundbreaking recognitions in the Ecuadorian Constitution of 2008 are the result of the long-standing debates of Latin American social movements and international-related networks. Several of the key actors of these debates participated in the Constituent Assembly. The enactment of *buen vivir* (living well) as the axis to policymaking and the making of the first constitution worldwide to recognise the rights of nature were made possible through the involvement of social movements. *Buen vivir* underpins a good life in equilibrium and coexistence with our habitats instead of aiming to improve living standards at the expense of other species and people (like those in extraction areas, for instance). Article 71 recognises the right of Nature (Pachamama) to exist, regenerate and fulfil its cycles, as an intrinsic value, independently of human needs. Both these concepts are complementary. They take human beings out of the centre and instead bring to the fore the plural socio-natural interactions, which are not exclusive to the Global South, as the only path to build a sustainable system. Also, urban manifestations and protests against fossil fuel impacts and their slogans contribute to raise awareness in new generations about the hidden costs of fossil fuels extraction. "Matar por petróleo, no es desarrollo" ("Killing for oil is not development") is one of the slogans heard in Quito during protests where urban groups show solidarity to the oil extraction problems in the Amazon. *Buen vivir* and the rights of nature are constitutional enactments that comprise the elements social movements often refer to in LFFU struggles. The Sápara nationality, at the brink of extinction, launched their own life plan, called *Kamunguishi*, which means "World." They claim the need for LFFU to preserve their territory and spirituality that make the forest they call "home." Sarayaku Kichwa communities have designed a life plan called *Kawsay Sacha* or "Living Jungle," free of extractive industries, where all beings of the jungle are respected and the rights of Indigenous nationalities to territory and Pachamama are recognised (Sarayaku, n.d.; Radcliff, 2012; Quick & Spartz, 2018). Paradoxically, the institutionalisation of these notions gave the Ecuadorian state a greener image, but this also served as a green façade behind which extractive projects not only continued but also additional oil blocks were opened in the Amazon while metal mining was promoted as a new growth sector (Lu et al., 2017; Valladares & Boelens, 2017; Van Teijlingen & Hogenboom, 2016). However, the enactment is compelling evidence that society is pushing to break the status quo and transform the way we envision well-being and coexistence with nature.

In the 2010s the region saw an explosion of initiatives that addressed the energy transition and its complexities. Latin American experiences and debates have sparked quests for an energy transition that go to the root of questioning the development model and our very relation with nature. In 2010 Bolivia organised The World People's Conference on Climate Change and the Rights of Mother Earth. The conference hosted social movements from all over the world. They prepared a declaration presented at the UN Climate Summit later that year. Despite the insurmountable contradictions of the experiences of Ecuador and Bolivia—at different levels, both countries had included *buen vivir* and the rights of nature in laws or official policy, while deepening fossil fuel extraction. Activists and renowned figures like Vandana Shiva and Nnimmo Bassey filed a lawsuit against BP in an Ecuadorian court for the *Deepwater Horizon* oil spill in the Gulf of Mexico in 2010. They did it to vindicate the rights of nature, particularly the rights of the sea, under the principle of universal jurisdiction. Meanwhile, a court in Colombia recognised the Atrato River and the Amazon as a subject with rights. In Argentina, a senator proposed a bill on the rights of nature, but it did not pass. Related to the mobilisation to rethink development and nature, social organisations have also been active in ways to rethink energy. Social movements demand democratic, decentralised, participatory and socially controlled access to energy. In 2019, over 10 social organisations of the continent issued the declaration

"In Latin America a new world is beating and energy will be its heart." The declaration points out the roots of extractivism, racism, patriarchy and coloniality in the current energy model. They call for a transition built from a popular project that guarantees energy as a human right, labour rights, participation to reflect people's needs and a socio-ecological transition. They make clear that "to change the energy model means to change the economic and social model." Lastly, the organisations celebrate the experiences of local communities, peasants, women and Indigenous people who carry out alternative energy processes, which are seedbeds for a new model (Soler, 2019, p. 31).

La Ruta Amazónica de la Esperanza (The Amazon Route of Hope) is one of these seedbeds. It takes place in the Ecuadorian north Amazon where a big part of the population depends on precarious working conditions and compensation programmes from the oil companies. People affected by oil extraction have worked for years to recover environmental, personal and communitarian health, and to find alternatives that give them economic autonomy and repair the damages related to oil extraction. This followed a process of recovering and re-discovering the equilibrium of the Amazon soil for food availability for people and animals; expanding artisanal production; small-scale energy generation with bio-digestors and biking machines; addressing issues of family and community violence and legal processes to hold companies accountable for their liabilities.

This is one of the many recent transition initiatives taking place at the local level throughout the Latin American region. Although the influence of LFFU social mobilisation in policy change varies and in part depends on the political and economic interests and circumstances, civil society is more aware and vigilant about fossil fuel extraction activities. The existing networks contest the industry's expansion at local and regional levels, while the increasing global push to LFFU and halt climate change inspires these networks (and vice versa). Their role is key in the race to stop the climate crisis, guide an energy transition that improves the quality of life for current and future generations and reduces the big inequalities that characterise Latin America. And under certain conditions, their proposals have a chance to become policy, such as recently in Colombia, where in 2022 an environmentalist politician—Gustavo Petro—became president after campaigning with a wide coalition of social movements, including trade unions, Afro-Colombian, Indigenous and women's movements and grassroots organisations. Phasing out oil is among the priorities of President Petro's government, and if it succeeds Colombia would be world's first major oil producing country to do so.

4.5. Conclusion

Since the turn of the century, social mobilisation for LFFU in Latin America has broadened and strengthened in several ways. First, various socio-environmental justice groups and agendas have come to criticise the dominant policies and practices of fossil fuel extraction, production and consumption. Next to the growing scientific evidence and tangible consequences of climate change, this shift resulted from the regional rise of Indigenous and peasant movements and ecofeminist currents. This has led to promoting niches demanding change. Second, in their local and national resistance against fossil fuel projects, some movements have managed to convince or force public sector institutions and other actors to become an ally in LFFU proposals. By mobilising state and other actors, several non-extraction initiatives such as moratoriums have been turned into concrete policy proposals, which have been adopted locally. This has led to changes in policy regimes. Third, beyond fossil fuel and LFFU, the long-standing critical debate about the region's commodity-based economic model being a source of dependency, inequality and underdevelopment has been revived with the insertion of socio-ecological concerns. This has mobilised profoundly new ideas about development and social and environmental rights. In a few recent cases, views on alternatives for development and on the rights of nature have changed public attitudes towards fossil fuel extraction and have become part of new legislation. Such changes are critical for changing landscapes. Still, even within strong multi-actor coalitions, it will be important to develop and implement a comprehensive strategy that connects such eco-inclusive approaches with socio-inclusive approaches at both the local and national scale (see Figure 4.3).

Progress towards LFFU driven by social movements remains fragile and contested and it can easily slide back when political shifts and economic crises occur. The case of Ecuador's Yasuní–ITT Initiative is the most prominent example of this risk of LFFU policy backsliding. However, the popular outcry and social mobilisation that followed also show that once new LFFU approaches and arguments have been developed and put in practice, they create new social and political supporters that are willing to take these ideas to the next level. The historical idea that fossil fuel reserves are a wealth and that they have to be exploited—the extractive imperative—are quite persistent as some political reactions to the recent (pre-pandemic and pandemic) economic problems showed. Still, in this ongoing process of the politics of ideas, resource nationalism around fossil fuels is losing ground, and leaving fossil fuels underground increasingly finds support from both the bottom and the top.

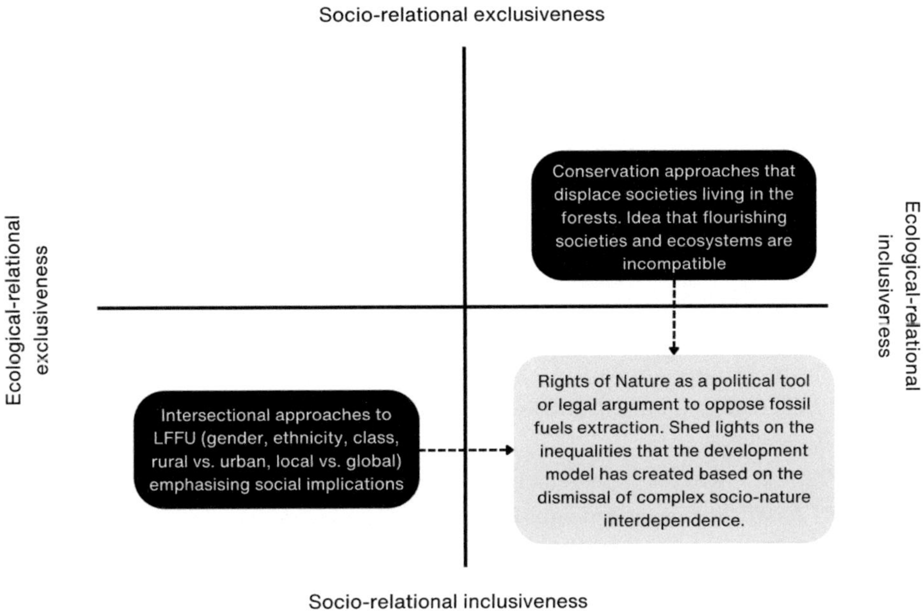

Figure 4.3. Latin America, social movements and inclusive transitions.

References

Acosta, A. (2013). Las tribulaciones de la Iniciativa Yasuní–ITT. *Letras verdes. Revista latinoamericana de estudios socioambientales, 6*, 19–22. https://doi.org/10.17141/letrasverdes.6.2010.871

Amazon Watch. (2018). Mandate from Amazonian women grassroots defenders of the rainforest against the extractive economic model. https://amazonwatch.org/assets/files/2018-03-mandate-from-amazonian-women-grassroots-defenders.pdf

Arsel, M., Hogenboom, B., & Pellegrini, L. (2016). The extractive imperative in Latin America. *The Extractive Industries and Society, 3*(4), 880–887.

Åsedotter Strønen, I. (2017). *Grassroots politics and oil culture in Venezuela: The revolutionary petro-state*. Springer Nature.

Bebbington, A. (2013). Natural resource extraction and the possibilities of inclusive development: Politics across space and time. Working Paper 21. Effective States and Inclusive Development Research Centre (ESID). https://www.effective-states.org/working-paper-21/

Bebbington, A., Abdulai, A. G., Humphreys Bebbington, D., Hinfelaar, M., & Sanborn, C. (2018a). *Governing extractive industries: Politics, histories, ideas*. Oxford University Press. https://doi.org/10.1093/oso/9780198820932.001.0001

Bebbington, D. H., Verdum, R., Gamboa, C., & Bebbington, A. J. (2018b). The infrastructure–extractives–resource governance complex in the Pan-Amazon. *European Review of Latin American and Caribbean Studies/Revista europea de estudios latinoamericanos y del Caribe, 106*, 183–208. https://doi.org/10.32992/erlacs.10414

Cardoso, A. (2015). Behind the life cycle of coal: Socio-environmental liabilities of coal mining in Cesar, Colombia. *Ecological Economics, 24*(120), 71–82. https://doi.org/10.1016/j.ecolecon.2015.10.004

Caretta, M. A., & Zaragocin, S. (2020). Women's resistance against the extractive industry: Embodied and water dimensions. *Human Geography, 13*(1), 3–5. https://doi.org/10.1177/1942778620910893

Castro, M. (2021, January 26). El colectivo Yasunidos denuncia a los exconsejeros del CNE 2014. *GK*. https://gk.city/2021/01/26/yasunidos-denuncia-consejeros-cne-2014/

CEJIL [Centro por la Justicia y el Derecho Internacional]. (2012, July 25). Inter-American Court condemns Ecuador for violating rights of Indigenous people of Sarayaku. https://cejil.org/en/press-releases/inter-american-court-condemns-ecuador-for-violating-rights-of-indigenous-people-of-sarayaku/

Christel, L. G., & Mariano, N. (2018). Incentivos económicos y conflictividad social: Trayectorias disímiles del fracking en las provincias argentinas (Neuquén y Entre Ríos, 2010–2017). *POSTData: Revista de reflexión y análisis político, 23*(2), 491–525.

Cielo, C., Coba, L., & Vallejo, I. (2016). Women, nature, and development in sites of Ecuador's petroleum circuit. *Economic Anthropology, 3*, 119–132. https://doi.org/10.1002/sea2.12049

Dinerstein, A. C. (2015). *The politics of autonomy in Latin America: The art of organising hope*. Palgrave Macmillan.

EJES [Enlace por la justicia energética y socioambiental]. (2018, July). Polémico fallo contra ordenanza que prohíbe el fracking en una localidad rionegrina. *Boletín* 20. https://www.ejes.org.ar/boletines/boletin20.html

El Universo (2013, October 20). Una lucha amazónica con rostro de mujer. *El Universo*. https://www.eluniverso.com/noticias/2013/10/20/nota/1603276/lucha-amazonica-rostro-mujer

Escobar, A. (2010). Latin America at a crossroads: Alternative modernizations, post-liberalism, or post-development? *Cultural Studies, 24*(1), 1–65. https://doi.org/10.1080/09502380903424208

Gudynas, E. (2010, April). Si eres tan progresista ¿Por qué destruyes la naturaleza? Neoextractivismo, izquierda y alternativas. *Ecuador debate. Conflictos del extractivismo, 79*, 61–82.

Gudynas, E. (2019). *Extractivismo y corrupción: Anatomía de una íntima relación*. Abya Yala.

Hernández Castillo, R. A. (2010). The emergence of Indigenous feminism in Latin America. *Signs: Journal of Women in Culture and Society, 35*(3), 539–545.

Hogenboom, B. (2017). Chinese influences and the governance of oil in Latin America the cases of Venezuela, Brazil, and Ecuador. In M. P. Amineh & G. Yang (Eds.), *Geopolitical economy of energy and environment: China and the European Union* (pp. 172–211). Brill.

IACHR [Inter-American Court of Human Rights]. (2010). Technical data: Pueblo Indigena Kichwa de Sarayaku vs Ecuador. https://www.corteidh.or.cr/cf/jurisprudencia2/ficha_tecnica.cfm?lang=en&nId_Ficha=206

IACHR [Inter-American Court of Human Rights], UN Human Rights & REDESCA. (2022, April 22). We must urgently protect environmental defenders. https://www.oas.org/en/IACHR/jsForm/?File=/en/iachr/media_center/PReleases/2022/084.asp

Jiménez, I. (2020, October). "*Fracking* ni hoy ni nunca": Entrevista a Tatiana Roa. Análisis y debate, no. 15. Rosa Luxemburg Stifting, Oficina Región Andina. https://www.rosalux.org.ec/pdfs/FrackingNiHoyNiNunca.pdf

Kimerling, J. (2006). Transnational operations, bi-national injustice: ChevronTexaco and Indigenous Huaorani and Kichwa in the Amazon rainforest in Ecuador. *American Indian Law Review, 31*(2), 445–508. https://doi.org/10.2307/20070795

Kothari, A., Salleh, A., Escobar, A., Demaria, F., & Acosta, A. (Eds.). (2019). *Pluriverse: A post-development dictionary.* Tulika Books and Authorsupfront.

Lang, M., & Mokrani, D. (2011). *Más allá del desarrollo.* Abya Yala and Fundación Rosa Luxemburgo.

Larrea, C. (2022). *Medio siglo de extracción petrolera en el Ecuador: Impactos y opciones futuras. Versión preliminar.* Universidad Andina Simón Bolívar.

Larrea, C., & Warnars, L. (2009). Ecuador's Yasuní–ITT Initiative: Avoiding emissions by keeping petroleum underground. *Energy for Sustainable Development, 13*(3), 219–223. https://doi.org/10.1016/j.esd.2009.08.003

Lu, F., Valdivia, G., & Silva, N. L. (Eds.). (2017). *Oil, revolution, and Indigenous citizenship in Ecuadorian Amazonia.* Palgrave Macmillan.

Lyall, A. (2018). A Moral Economy of Oil: Corruption Narratives and Oil Elites in Ecuador. *Culture, Theory and Critique*, 59(4), 380–399.

Martinez-Alier, J., Baud, M., & Sejenovich, H. (2016). Origins and perspectives of Latin American environmentalism. In F. de Castro, B. Hogenboom & M. Baud (Eds.), *Environmental Governance in Latin America* (pp. 29–57). Palgrave Macmillan.

Moreano, M. (2014). Ecuadorian anti-extractivism: Limits and possibilities of a convergence between the ecologists and the Indigenous movements. Working paper. Society for Latin American Studies 2014 Conference (SLAS) Birkbeck, University of London, 3–4 April 2014.

OPSUR [Observatorio Petrolero Sur]. (2016, December 26). Un año de victorias en la lucha contra el fracking. https://opsur.org.ar/2016/12/26/un-ano-de-victorias-en-la-lucha-contra-del-fracking/

Pellegrini, L. (2018). Imaginaries of development through extraction: The "History of Bolivian Petroleum" and the present view of the future. *Geoforum*, 90, 130–141. https://doi.org/10.1016/j.geoforum.2018.01.016

Peters, S. (2017). Beyond curse and blessing: Rentier society in Venezuela. In B. Engels & K. Dietz (Eds.), *Contested extractivism, society and the state: Struggles over mining and land* (pp. 45–68). Palgrave Macmillan.

Plataforma Latinoamericana. (2021, May 7). Declaración de la Plataforma Latinoamericana y del Caribe por la Justicia Climática frente a la Semana Regional del Clima. https://agenciaecologista.info/2021/05/07/declaracion-de-la-plataforma-latinoamericana-y-del-caribe-por-la-justicia-climatica-frente-a-la-semana-regional-del-clima/

Quick, J., & Spartz, J. T. (2018). On the pursuit of good living in highland Ecuador: Critical Indigenous discourses of Sumak Kawsay. *Latin American Research Review*, *53*(4), 757–769.

Radcliffe, S. A. (2012). Development for a postneoliberal era? Sumak Kawsay, living well and the limits to decolonisation in Ecuador. *Geoforum*, *43*(2), 240–249.

Reuters. (2021, January 25). European lenders exit Amazon oil trade after scrutiny by campaigners. https://www.reuters.com/business/energy/european-lenders-exit-amazon-oil-trade-after-scrutiny-by-campaigners-2021-01-25/

Ross, M. L. (2012). *The oil curse: How petroleum shapes the development of nations*. Princeton University Press.

Santiago, M. I. (2006). *The ecology of oil: Environment, labor, and the Mexican Revolution, 1900–1938*. Cambridge University Press.

Sarayaku. (N.d.). Propuesta del pueblo originario Kichwa de Sarayaku. https://sarayaku.org/en/propuesta-kawsak-sacha/?lang=es

Science Panel for the Amazon. (2021). The Amazon we want. https://www.theamazonwewant.org

Serje, M. (2003). ONGs, indios y petróleo: El caso U'wa a través de los mapas del territorio en disputa. *Institute Francais d'Etudes Andines, Peru*, *32*(1), 101–131. https://doi.org/10.4000/bifea.6398

Serrano, D. (2022, September 8). Moratoria petrolera se mantendrá hasta establecer consulta previa. *El Comercio*. https://www.elcomercio.com/actualidad/negocios/moratoria-petrolera-mantendra-hasta-consulta-previa.html

Silva, E., Akchurin, M., & Bebbington, A. (2018). Policy effects of resistance against mega-projects in Latin America. *ERLACS: European Review of Latin American and Caribbean Studies/Revista europea de estudios latinoamericanos y del Caribe*, *106*, 25–46.

Soler, J. P. (2019). *Transición energética en América Latina. Propuestas de transición energética en America Latina: Agua y energía para nuestra soberanía.* CENSAT.

Svampa, M. (2015). Commodities consensus: Neoextractivism and enclosure of the commons in Latin America. *South Atlantic Quarterly, 114*(1), 65–82. https://doi.org/10.1215/00382876-2831290

TCE [Tribunal Contencioso Electoral]. (2022, September 5). Sentencia causa no. 888-2019-TCE. https://www.yasunidos.org/wp-content/uploads/2022/09/SENTENCIA_CAUSA_Nro.888-2019-TCE.pdf

Temper, L., Yánez, I., Sharife, K., Godwin, O., & Martinez-Alier, J. (Eds.). (2013, May). *Towards a post-oil civilization: Yasunization and other initiatives to leave fossil fuels in the soil.* EJOLT report no. 6. http://www.ejolt.org/wordpress/wp-content/uploads/2013/05/130520_EJOLT6_Low2.pdf

Time. (2020). The 100 most influential people of 2020. https://time.com/collection/100-most-influential-people-2020/5888337/nemonte-nenquimo/

Tudela, F. (2020). Obstacles and opportunities for moratoria on oil and gas exploration or extraction in Latin America and the Caribbean. *Climate Policy, 20*(8), 922–930.

Valdivia, G., & Lyall, A. (2018). The oil complex in Latin America: Politics, frontiers, and habits of oil rule. In J. Cupples, M. Palomino-Schalscha & M. Prieto (Eds.), *The Routledge handbook of Latin American development* (pp. 458–468). Routledge

Valladares, C., & Boelens, R. (2017). Extractivism and the rights of nature: Governmentality, "convenient communities" and epistemic pacts in Ecuador. *Environmental Politics, 26*(6), 1015–1034. https://doi.org/10.1080/09644016.2017.1338384

Van Teijlingen, K., & Hogenboom, B. (2016). Debating Alternative development at the mining frontier: *Buen vivir* and the conflict around El Mirador mine in Ecuador. *Journal of Developing Societies, 32*(4), 382–420. https://doi.org/10.1177/0169796X16667190

Vogue. (2021). 7 inspiring young activists working to save the planet. https://www.vogue.co.uk/news/article/climate-activists-on-instagram

Yashar, D. J. (2005). *Contesting citizenship in Latin America: The rise of Indigenous movements and the postliberal challenge.* Cambridge University Press.

5. South Africa and the Political Economy of Fossil Fuels: Rationales for LFFU

Patrick Bond and Arthur Rempel

Abstract

South Africa has relied on extremely high levels of coal and other fossil fuels, but the early 2020s was a period of flux in South Africa. Dominant patterns of fossil dependency began to change and the politics of ideas shifted not only to renewable energy, but also to an explicit commitment made by the government to LFFU. This chapter first looks into South Africa's national context vis-à-vis climate policy and the related key actors. Next it discusses the key arguments and approaches for leaving fossil fuels underground (LFFU) that would also rectify the injustices of apartheid capitalism and address climate costs. Against a broader context of high-level corruption and collusion, the deep-rooted minerals–energy complex and the related lack of climate action, LFFU approaches that are being deliberated in South Africa include stranded assets, vulnerabilities to international trade, extractivism accounting, climate debt and the social cost of carbon (SCC). South Africa serves as a vital case of considering the politics of ideas that play an important role in changing dominant patterns.

Keywords: fossil fuel dependency, political economy, injustices, climate costs, South Africa

5.1. Introduction

South Africa has relied on such extremely high levels of coal and other fossil fuels for energy and export earnings, that it has been responsible for 29 billion tonnes of CO_2-equivalent (CO_2e) emissions from 1850 to 2021 (Evans, 2021). This amount still grows by 500 million tonnes annually, which

Gupta, J., Hogenboom, B., Rempel, A. & Olofsson, M. (eds), *Leaving Fossil Fuels Underground: Actors, Arguments and Approaches in the Global South and Global North*.
Taylor & Francis Group, 2024
doi: 10.5117/9789048560370_CH05

if divided by 60 million people, and again divided by a $420 billion current GDP, would rank third among countries of at least 10 million people (only Kazakhstan and the Czech Republic are higher). This per capita annual emission of 8.3 tonnes of CO_2e is due not to the process of industrialisation, but because of carbon-intensive mega-project development of an *extractive* character, mainly in deep mining, smelting, petrochemicals and other capital-intensive sectors with few backward–forward linkages. Obviously, there is a race–class–gender–location factor. The carbon footprint of white, wealthy, urban males (and the companies they run or that they invest in) is far higher than that of black, poor, rural women, in what is the world's most income-unequal country.

This legacy naturally generates fierce debates about energy justice. It motivates us to provide background on why South Africa has such a fossil-addicted economy, and to discuss strategies for leaving fossil fuels underground (LFFU) that would rectify the injustices of apartheid capitalism as well as address climate costs. The national policy context is vital, as are the central actors who continue to rely upon not only a coal-intensive regime, but also an emerging methane gas economy, within what is known as South Africa's minerals–energy complex (MEC) (Fine & Rustomjee, 2018) (see 5.3). The struggle to shut down coal-fired power plants and coal mines is fraught with danger (e.g. a high-profile activist's assassination in late 2020), but to some extent provides hope given Western incentives to leave coal unmined through the $8.5 billion Just Energy Transition Partnership (JETP). Climate activists have also, since late 2021, adopted a militant stance against offshore gas exploration and liquefied natural gas energy generation. They began not only protesting weekly on the beaches and taking firms exploring for gas to court (winning anti-exploration injunctions in six out of the first seven cases through mid-2022), but also potentially leveraging South Africa's trade vulnerabilities given the likelihood of climate sanctions, and more openly accounting for extractivist harms (e.g. by invoking the social cost of carbon [SCC]), abandoning GDP fetishisation (since pollution and depletion are neglected), and addressing South Africa's climate debt (see 5.4). These progressive arguments have been met with resistance from the incumbent regime and Big Oil and Coal (see 5.5).

5.2. Building on the analytical framework

Uniquely in South Africa, there is space to consider three features that build on the LFFU analytical framework (see Chapter 2): (1) encouraging the prospect of "climate sanctions" (the Carbon Border Adjustment Mechanism) against South African exports that have a high embedded energy content drawn from coal or gas; (2) the SCC as a mode of encouraging stranded-asset analysis; and (3) generational justice in contemplating non-combustible use of hydrocarbons (hence drawing in "natural capital accounting" in a manner that can halt current fossil fuel extraction rather than encourage it).

To assess how these arguments relate to prevailing wisdom about LFFU and the South African policy context, we take as given the arguments in Chapter 2 in relation to the comprehensive multidimensional perspective required for inclusive development, as well as North–South dimensions and the global vision of climate justice movements. Hence, we move beyond what colleagues term the limited "supply-side process of moving from high-GHG-emitting energy sources to low- to no-GHG-emitting energy sources" which in turn "means moving from coal to oil to gas to renewables" (Chapter 2). South Africa illustrates the urgency of changing the terrain of debate so as to highlight "the demand side of infrastructure and technology" in order that the full range of socio-ecological-economic costs of fossil fuels are accounted for. To that end, South Africa is a vital case of considering "the *politics* of ideas that play an important role in changing dominant patterns," as proposed in Chapter 2.

The early 2020s was a period of flux in South Africa, where indeed dominant patterns of fossil dependency *did* begin to change and the politics of ideas shifted to not only renewable energy, but to an explicit commitment made by the government to LFFU. This occurred in 2021, with Pretoria's statement of intent to the United Nations to pursue "non-fossil development in Mpumalanga" (the province that was South Africa's traditional coal zone), and the commitment by the CEO of South Africa's electricity utility Eskom (André de Ruyter) to more rapidly close coal-fired power plants. Both were conditional on receiving concessional finance, and to that end in October 2021, just before the Glasgow climate summit, Western governments pulled together $8.5 billion in credit lines to finance decarbonisation (of which the French and German components were substantially below market while US and British loans were at market rates). Hence there was a brief period of 14 months when Western "concessional" finance to incentivise LFFU appeared as a genuine policy and even practical opportunity. Indeed at surface level, South Africa's LFFU strategy appeared to conform to many of the components that *could* be considered steps toward compensation of Southern countries: first, genuinely recognising fossil fuels as stranded assets and committing not to extract them; second, ensuring that a "just transition" accompaniment exists for fossil fuel-dependent workers and communities (hence diminishing their opposition to change); and third, gaining finance from taxpayers in the Global North. That was the theory that made the JETP appear consistent with inclusive development, but the weight of tradition and the logic of MEC power relationships prevented its realisation.

5.3. Contextualising South Africa's fossil fuel regime

5.3.1. Key policy documents

Two policy documents are relevant for our analysis of South Africa, aside from erratic JETP documentation: the Integrated Resources Plan (IRP) of 2019 (which prevailed as this book went to press in 2023), and South Africa's Nationally Determined Contribution submission to the Paris Climate Agreement of 2015, updated for the Glasgow COP26 in 2021 and the Sharm El Sheikh COP27 in 2022.

The Integrated Resource Plan
South Africa's Integrated Resource Plan (IRP) (2019) is the national electricity plan. After enormous influence was exerted by the Energy Intensive Users Group of Southern Africa (EIUG)—the main multinational extractive industry corporations' lobby of 27 firms that use 40% of the country's electricity—the IRP that set the tone for high fossil-based energy use (instead of renewables) in recent years, was published in 2010. The 2019 revision continued to suggest a role for new coal-fired power plants as well as methane gas and nuclear energy generation (DMRE, 2019), all of which are opposed by climate activists. Even the Presidential Climate Commission opposed coal and nuclear, but also suggested 3,000 to 5,000 MW of new "peaking" gas to support renewables, whereas activists instead supported pumped hydropower storage and molten salt at solar chimney sites as relatively non-invasive solutions to periods of inadequate renewable supply. While the IRP establishes the schedule by which South Africa will install new power capacity through 2030 to meet energy demands, it requires yet another revision given the changing global and local conjunctures, including South Africa's rapidly worsening energy poverty and insecurity, including major electricity cuts and prices soaring 10% in real terms.

As of 2018, South Africa's installed electricity capacity stood at roughly 51.6 GW, of which 37 GW (72%) was coal-based and merely 3.5 GW (7%)

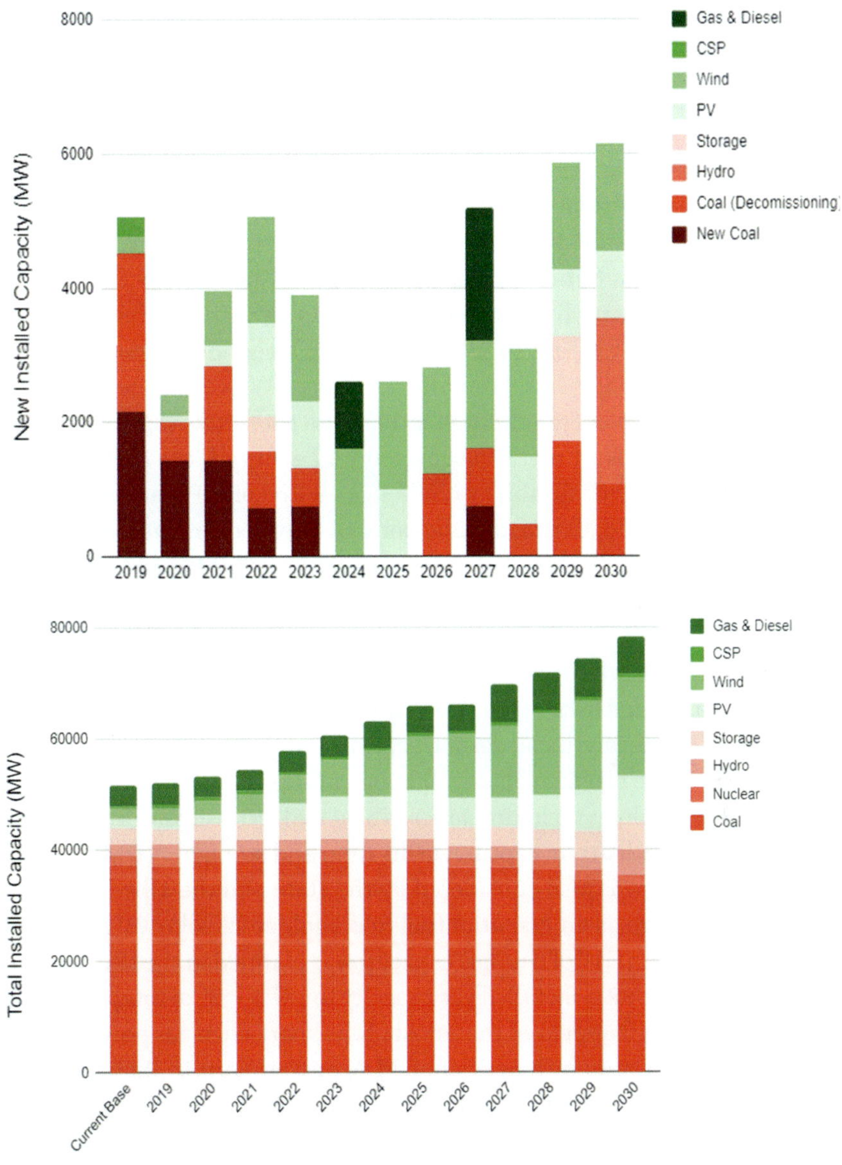

Figure 5.1. Energy projections for South Africa's new (top) and total (bottom) installed capacity by 2030.
Source: Rempel (2023).

was solar PV and wind power, and the rest a combination of nuclear (Cape Town's Koeberg reactor, the only such source in Africa), hydropower (from the Cahorra Bassa dam in Mozambique) and other imports from the regional grid. According to the IRP, South Africa will decommission

11 GW of coal capacity by 2030, but simultaneously procure and install 7.2 GW of new coal, yielding an installed coal capacity of 33.3 GW by 2030 (DMRE, 2019), thus decreasing South Africa's total coal capacity by slightly over 10%. Total installed capacity across all energy types is projected to increase by 27 GW (over 50%) to 78.2 GW, which will be met by investments in new solar (7 GW), wind (16 GW), gas and diesel (3 GW) and hydro (2.5 GW) installed capacity. Although these projections do indicate a diversification in South Africa's energy mix—namely by decreasing coal installed capacity from 72% of the total to 43% and increasing solar PV and wind power shares to 11% and 23%, respectively—the aggregate installed coal capacity would remain at over 33 GW over the next decade, hinting at a possible energy *addition* rather than *transition,* especially given Eskom's drive to insert methane gas-sourced energy (York & Bell, 2019). Moreover, due to the variability challenges with solar PV and wind, the IRP itself acknowledges that coal power will likely contribute to meeting 59% of annual energy demand (in MWh), while solar PV and wind will jointly provide only 24%.

Through the IRP, South Africa's Department of Mineral Resources and Energy (DMRE) clearly envisages coal playing "a significant role in electricity generation in South Africa in the foreseeable future" and underscores the importance of (largely mythical) carbon capture and storage schemes, through investing in "more efficient coal technologies (HELE [high-efficiency, low-emissions] technology, including supercritical and ultra-supercritical power plants with CCUS [carbon capture, utilisation and storage])" (DMRE, 2019, p. 12). Methane gas is also considered by the state and the allied National Business Initiative as an essential fuel to complement solar PV and wind power in the mix; 3 GW of new gas capacity was planned in the IRP through 2030, bringing the total installed gas capacity to 7 GW (9% of total capacity). By 2023, however, it appeared that most of the short-term gas power would be generated not only by a new 3,000 MW power plant in Richards Bay, but also by three Turkish floating Karpowerships LNG generators to be placed in the Richards Bay, Coega and Saldanha harbours. Allegations of Karpowership's corruption of government officials (made even by Eskom's CEO De Ruyter in 2023), of local marine ecology damage and of the ships' methane contributions to the climate crisis slowed but apparently would not stop the ships from being ordered.

A final factor of enormous importance to the country's renewable energy capacity is that the *transmission* of this power has been skewed by the location of grid access in the eastern part of the country in the vicinity of a dozen major coal-fired power stations. The optimal solar and wind facilities are on the coastal and especially semi-desertified western areas, where there is minimal transmission capacity due to the lack of foresight of the electricity utility Eskom. The capacity for transmission which does exist was quickly saturated by the large mining and smelting firms attempting to lower their fossil fuel reliance during the early 2020s. Hence, they "wheeled in" renewable energy so as to prepare for a period in which a version of climate sanctions will apply to their output, as the West adopts carbon border adjustment taxes to penalise overseas suppliers with high carbon inputs embedded in their products.

The impact of sanctions on fossil fuels was indeed formidable. For example, although it was not mentioned in the IRP, South African President Cyril Ramaphosa arrived back from the Forum on China–Africa Cooperation in September 2018 and told his home province that they would benefit from a special economic zone (SEZ) (to be run by Chinese entrepreneur Ning Yat Hoi, who at the time was on the Interpol red list, see 5.3.2) that would be powered by a 4,600 MW coal-fired power plant. This energy source was shelved in early 2022, according to the project's South African manager, because of oppositional activism and Xi Jinping's September 2021 announcement that the Belt and Road Initiative would contain no further coal-fired power plants (Cronje, 2022). Financial sanctions against coal-fired power were imposed by most of the major lenders both locally and globally in the late 2010s. Another form of international name-and-shame sanctions occurred at a 2019 United Nations climate summit due to South Africa's lack of emissions-cutting ambition. And, as discussed below, the mid-2020s would witness Western trade sanctions against South African products that included embedded energy with a high CO_2 content were also important motivations to LFFU.

Nationally determined contribution[1]

The 2009 Copenhagen Accord spelled out a so-called "bottom-up" system for reducing GHG emissions, contingent upon each signatory country committing to make cuts on their own volition, with increasing ambition in each iteration. South Africa's nationally determined contributions (NDCs) (originally published in 2010 and updated in July 2021 by the Department of Forestry, Fisheries and the Environment [DFFE]) spelled out planned commitments to gradually align with the Paris Climate Agreement's 1.5°C aspirational target. The plan follows a "Peak–Plateau–Decline" logic proposing to peak greenhouse gas emissions between 2020 and 2025, "plateau for

1 This section is adapted from Rempel (2023).

approximately a decade and decline in absolute terms thereafter" (Republic of South Africa, 2021, p. 7). The initial draft proposed that its emissions would fluctuate between 398 and 614 Mt CO_2e between 2025 and 2030 before declining, but after critiques of this being a "highly insufficient" target and aligning with a 4 °C warming scenario (Climate Action Tracker, 2022), the updated NDCs lowered this upper limit by 28%, so now by 2030 South Africa proposes lowering its annual emissions to the 350–420 Mt CO_2 range. Climate Action Tracker classifies this new target as "insufficient" as it likely aligns with 3 °C warming and therefore is still incompatible with the Paris Agreement objectives in spite of making an improvement to the first iteration (Climate Action Tracker, 2022).

However, while South Africa's NDCs were itself inadequate, one extremely important proposal was made: "The just transition in South Africa will require international cooperation and support ... by the international climate and development and finance community for non-fossil-fuel development in Mpumalanga" (Republic of South Africa, 2021, p. 28). In March 2021, Environment Minister Barbara Creecy claimed that concessional finance should be paid to South Africa to incentivise leaving coal underground—even while gas and oil were being rapidly sought through offshore exploration and onshore fracking licences.

The strength of the NDCs is that they acknowledge that coal must be left underground, albeit as a way to gain some form of concessional financing from the Global North. However, there are alarming weaknesses in the language of the NDCs, including neglect of South Africa's myriad of fossil-intensive mega-projects, and its failure to attribute the highest carbon footprints to South African elites, especially those associated with the MEC. Given the MEC's centrality within a deindustrialising South African capitalism, and the rapidly worsening power cuts (termed "load shedding") starting in 2008 but even more rapidly from 2021, it would have been appropriate for the NDCs to commit to energy rationing on the basis of Scope 3 emissions, through which the MEC's contributions to climate change above and beyond consumption of Eskom's coal-fired power would be penalised.

Still, the creative suggestion for concessional finance to thereby avoid further Mpumalanga coal mining could have been a decisive way to generate emissions cuts and just transition strategies. The November 2021 offer of $8.5 billion by Western governments included only 3% grant funds, along with relatively low interest rates from Europe but market rates from the US and United Kingdom. This world-first "just energy transition partnership" fell short in many critical respects, including South Africa's leadership failure, which itself is correlated to systematic corruption of the state and electricity company Eskom—the country's largest state-owned enterprise.

5.3.2. Key actors

The most striking feature of the way powerful actors have addressed fossil fuels and LFFU is how much corruption has been unveiled in the process. Ramaphosa's role is revealing because he was previously a coal company tycoon until becoming deputy president in 2014, and his own firm (Shanduka) worked closely with transnational coal and extractive companies such as Glencore and Lonmin (Bond, 2018). The 2018–2022 Zondo Commission investigating corruption heard testimony from a former CEO of Eskom that when Ramaphosa raised the price of coal purchased from the Glencore-Shanduka Optimum mine (which he had chaired until mid-2014) by a factor of more than four times—when as deputy president he ran the "Eskom War Room" in 2014–2015—he was actually acting on behalf of Switzerland-based Glencore. That firm was successfully prosecuted in 2022 for extensive bribery of African governments (although South Africa was not mentioned).

As mentioned above, in 2018 Ramaphosa promised a $27 billion metallurgical special economic zone (SEZ) for his home province of Limpopo, powered specially by a massive coal-fired plant, but, as noted earlier, the main Chinese entrepreneur was wanted by Interpol for mining-related corruption at the time.[2] In July 2021, President Ramaphosa restated a 2012 strategy within the National Development Plan (NDP), which he co-chaired, to export 18 billion tonnes of coal from his home province thanks to generous subsidisation: "The new mining developments in the Waterberg in Limpopo depend on the state for efficient and low-cost transport and energy" (SA News, 2021). The new Waterberg coal mines, rail line and coal-bearing locomotives were meant to be South Africa's largest-ever mega-project,

2 By 2021 the plant's proposed size was reduced to 1,320 MW and several of the anticipated smelters were downgraded or eliminated due to environmental, conservation and community resistance. But the massive adverse implications for South Africa's NDC were not mentioned. There was in 2020–2021 an attempt to mitigate CO_2 emissions by reference to a notional carbon capture storage system for sequestering the CO_2 underground and further SEZ plans even suggested nuclear energy would be used, so that the vast carbon-intensive industrial process could proceed. Ultimately a solar power plant was proposed in 2023, but it too could not disguise the carbon-intensive character of the mining, smelting, mineral processing and industrial manufacturing proposed for the SEZ, amounting to an additional 8% of South Africa's upper limit of 2030 emissions.

estimated at $53 billion (Govender, 2013). More than $1 billion in corruption on the Chinese locomotives was discovered in 2017 (Ensor, 2023). In 2020, also in the Waterberg region, a lucrative game farm wholly owned by Ramaphosa was the site of a robbery of millions of US dollars that he had stored in the cushions of a couch. After 2022 revelations about the illicit funds and alleged torture of the suspects, he was nearly forced to resign later that year, but his African National Congress (ANC) parliamentary allies rejected the findings of an official investigation, so he kept his job (Haffajee, 2023). In short, this was an extremely unlikely leader to genuinely decarbonise South Africa.

There are also numerous other cases of corruption in the fossil fuel sector that help explain reluctance to LFFU. Senior Eskom officials—including the chairperson at the time (2005–2008), Valli Moosa—were involved in Hitachi's bribery of the ruling ANC, when commissioning coal-fired power plant boilers worth billions of dollars in 2007–2010. Moosa sat on the ANC finance committee at a time Hitachi gave the party's investment arm 25% of its local subsidiary as a gift. In 2009, the public protector judged Moosa's conflict of interest to be "improper," yet he was still made the leader of Ramaphosa's Presidential Climate Commission a decade later. Hitachi paid $19 million as a fine to the US Securities and Exchange Commission under the Foreign Corrupt Practices Act in 2015. But as De Ruyter (2023, p. 148) pointed out,

> Shockingly, or perhaps not, this development was never investigated further by the South African authorities.... Medupi and Kusile soon became Eskom's terrible twins, wreaking havoc with its finances and operational efficiency. For the ANC's 97 million pieces of silver, the taxpayer is paying billions. Because of Kusile's design flaws, we've lost around 2,100 MW of generation capacity during 2023, enough to eliminate two stages of load shedding. To fix only some of the boiler mistakes will cost R4.2 billion, and we've already burnt R30 billion of diesel due to lost capacity. Measuring the indirect costs, like the loss of investors and job opportunities, is almost impossible. The decisions taken in 2007 will haunt South Africa for decades to come.

In 2023 there was no comment from Western government funders of the JETP when Ramaphosa, his electricity, energy and environment ministers and even the Presidential Climate Commission agreed that the timetable for retiring coal-fired power plants would be extended by years (as a result of the energy crisis). Commission executive director Crispian Olver openly admitted the state's lack of "capacity to prevent theft and corruption through the use of this money.... [W]e are emerging from state capture and very high corruption and what it's done to the basic fabric of the state and its ability to do stuff" (Leshoro, 2023). A few years earlier, Olver's own consulting firm Linkd was accused by South Africa's public protector of being the main vehicle for Ramaphosa's 2017 presidential campaign's "money laundering," amounting to what the president himself admitted was $15 million (Smit, 2019). That campaign received a grant of $145,000 from the CEO of the main local gas exploration firm, Impact Africa's Johnny Copelyn, and in spite of methane being 85 times more potent a greenhouse gas than CO_2, Ramaphosa lauded Copelyn and his partners in TotalEnergies—who in early 2019 located a major offshore condensate gas seam—for "a game-changer for our country. We are extremely encouraged" (SA News, 2018).

Meanwhile, Minister of Mineral Resources and Energy Gwede Mantashe was accused of accepting bribes from a company similar to the Gupta empire—Bosasa, run by the Watson brothers—as well as improper conduct in relation to what was originally a $15 billion energy contract with Karpowerships, tethered to three South African ports (Van Tilberg, 2023). As a former leader of the mineworkers union and as chair of the ANC, Mantashe regularly opposed both the closure of coal mines and the retiring of Eskom plants early. Yet he also blocked progress on a long-overdue just transition strategy to mitigate the damage to workers and communities. His failure to regulate extreme pollution associated with electricity generation and mining confirmed a lack of genuine concern for worker and community welfare. Accused of corruption by the Zondo Commission in relation to Bosasa personal bribery, he also accepted a $1 million contribution from Shell—into ANC coffers, as chair of the party in 2021 (via a party investment firm that owned 28% of the London firm's local affiliate)—at the same time he was approving Shell's offshore gas and oil explorations (Editor BizNews, 2023).

Environment Minister Creecy, acting in consort with Minister of Mineral Resources and Energy Mantashe, was often criticised for illogical approvals of environmental impact assessments associated with fossil fuels. For example, the Centre for Environmental Rights asked the minister to reduce SO_2 and NO_x emissions from power plants, refineries and other industrial facilities on the Mpumalanga Highveld (one of the world's worst pollution hotspots according to the 2007 National Air Quality Act), which affected 10,000 lives annually, and although the ministry drafted regulations, Creecy continually procrastinated in their implementation, as the two main culprits—electricity company Eskom and energy and chemical company Sasol (both South African)—were simply too influential (Bega, 2019). As De Ruyter (2023, p. 127) explained,

To fit the flue gas desulphurisation units is the power station equivalent of overhauling a car and stripping it down to its chassis. We simply do not have the time or the money. By appealing against the instruction from the national air quality officer, we were able to obtain a stay of execution while we made representations to a committee appointed by Minister Creecy.

Another important actor is the National Treasury, which imposed austerity on the poorest South Africans starting in late 2020 as the conditions of $4.3 billion from the International Monetary Fund (IMF) took effect but did not reduce the fossil fuel subsidies and resource gifts given to fossil fuel companies. Even staff at the IMF (Parry et al., 2021) accused the National Treasury of providing the industry with $51 billion in annual implicit subsidies, given its tokenistic $0.30/tonne carbon tax. The National Treasury also provided $70 million in mid-2021 to the South African National Defence Force to fund troops attempting to defeat Islamic rebels disrupting the world's third-largest gas field, in the Rovuma Basin offshore Cabo Delgado, Mozambique. The war threatened a $20 billion investment by TotalEnergies, so Emmanuel Macron in June 2021 visited both Paul Kagame and Ramaphosa, and within a few weeks, thousands of Rwandan and South African soldiers had joined the fighting on behalf of the fossil fuel industry, in spite of the increasingly powerful cyclones that were hitting central and northern Mozambique as the water in the Rovuma Basin regularly warmed to 30 degrees (Bond, 2022).

Leadership of Eskom has been highly unstable since the start of the 21st century. With De Ruyter (2023) at the helm, a degree of corruption by his predecessors and by ongoing crime syndicates was uncovered. De Ruyter did shift Eskom into at least a conceptual orientation towards renewable energy. But in addition to promoting methane gas (as Eskom closed down coal-fired power plants), he failed to either generate new renewable electricity or build transmission lines to the sunnier, windier parts of the country where private sector independent power producers (IPPs) were doing so. De Ruyter endorsed the privatised agenda that would not only split Eskom up and commercialise it, but also lead to higher prices than if the national utility were serious about a genuine just transition. As for the race and class politics of electricity supply, Eskom became infamous in the winter of 2020—in the worst period of COVID-19—when De Ruyter's new "load reduction" policy led to mass disconnections in black neighbourhoods and small towns where payment arrears rates were high, such as Soweto with a 60% non-payment record in 2022 (Pijoos, 2022), an explicit case of Eskom's new "energy racism" (Sinwell et al. 2022). De Ruyter resigned in late 2022 when he was accused of treason by Mantashe due to worsening power outages, and indeed his coffee was poisoned after he opposed crime syndicates and hired an investigations unit which apparently implicated at least two national ANC leaders (De Ruyter, 2023).

Civil society groups played diverse roles in challenging Eskom and others within the South African fossil fuel regime (see also Chapter 6). Notably, in 2020 in spite of COVID-19 lockdown restrictions, 350 Africa formed the Climate Justice Coalition (CJC), a campaign involving three dozen South African organisations which aspire to "overcome the power of polluters and politicians blocking action on climate justice" by building "a deeper, stronger and more powerful movement" (350 Africa, n.d.). CJC members range in size, including the 600,000-strong South African Federation of Trade Unions (SAFTU) (albeit with its largest member, the metalworkers, firmly adopting a pro-coal position in 2022), and community-based movements such as Mining Affected Communities United in Action (MACUA), the South Durban Community Environmental Alliance (SDCEA) and the Soweto Electricity Crisis Committee. The CJC's "Green New Eskom" campaign demands "a rapid and just transition to a more socially owned, renewable energy powered economy, providing clean, safe, and affordable energy for all, with no worker and community left behind in the transition" (CJC, 2021, p. 8). In addition, the Climate Justice Charter Movement (CJCM)—whose roots are in the cooperative movement but which promotes a high-level eco-socialist consciousness—petitioned Ramaphosa (unsuccessfully) to declare a climate emergency, and then in 2022 called for a Western boycott on funding Eskom due to the government's ongoing reliance on—and new exploration of—fossil fuels. In addition, broader-based just transition demands—e.g. the "Million Climate Jobs" campaign of the AIDC (Alternative Information and Development Centre) (2017)—were made regularly.

Among the private sector actors that exercised influence, the main collective agencies were the National Business Initiative, Business Unity South Africa and the historic pro-coal electricity consumers: the Energy Intensive Users Group of Southern Africa (EIUG). Its 27 members included the largest mining, smelting and electricity-guzzling firms in Africa, and included two companies—BHP Billiton (South32) and Anglo American—which from the early 1990s had enjoyed Eskom's special pricing arrangements: coal-fired electricity supply for smelters costing just \$0.01/kWh. By the early 2020s, that was just a 10th of what consumers paid and was a quarter of the price of electricity production.

5.4. Arguments and approaches to LFFU

Against this broader context of high-level corruption and collusion, the deep-rooted minerals–energy complex and the related lack of climate action, this section discusses some of the key arguments and approaches to LFFU that are being deliberated within the South African context.

5.4.1. Stranded assets

South Africa's economy developed a profuse dependence on cheap coal over the 20th century for both domestic consumption and as a source of export revenue (Baker et al., 2014; Swilling et al., 2016). In the late 2010s, 90% of South African energy and roughly 75% of primary energy demands were met with coal (Cock, 2019). The breakdown of coal-fired power plants, especially in 2022–2023, led hundreds of thousands of wealthier households to install personal solar systems (typically costing $10,000 each), which caused a slight decline in CO_2 emissions but also, adversely from the standpoint of energy justice, put pressure on municipalities which had used electricity revenues to cross-subsidise power supply and other services to poor residents. It often appears that large parts of the electricity grid will degenerate into "stranded" status.

While coal resources are abundant, oil and gas resources were until the late 2010s considered mainly absent in South Africa, aside from small gas deposits drawn into the Mossgas facility on the Indian Ocean during anti-apartheid oil sanctions. In part due to sanctions, a Fischer–Tropsch process (converting coal and later Mozambican gas into a synthetic fuel through gasification and liquefaction) was established by South African energy and chemical company Sasol at Secunda to meet 20% of its liquid fuel needs (Winkler & Marquand, 2007). But the prospect for a massive new gas-related infrastructure push intensified in 2019, when TotalEnergies made offshore discoveries in Exploration Block 11B/12B. Given the block's proximity to the shuttered Mossgas, which had run out of feedstock, the prospect of a new national pipeline system emerged with periodic claims that the methane gas could be added to South Africa's energy mix. Critics insist that exploring, extracting, processing, transporting and eventually combusting these resources would be in direct violation of the commitments made in the Paris Agreement and would also attract methane border taxes by Western trading partners. If the actual cost of emitting a tonne of carbon is $3,000 (as argued by Kikstra et al., 2021), then any net present value analysis would suggest a net negative value.

5.4.2. Vulnerabilities to international trade

Since South Africa exports heavily to Europe and the US, it is vulnerable to the expected Carbon Border Adjustment Mechanism (CBAM) from 2026 onwards. These exports traditionally were weighted towards luxury automobiles, smelted base and precious metals (especially steel, aluminium, platinum, gold, chrome and manganese), petrochemicals and some raw commodities (including iron ore), as well as agricultural produce (especially wine, horticulture and citrus) and some high-quality garments and other manufactured products. Not only the high-carbon autos and metals, but all products, could be subject to punishingly high tariffs on climate grounds (Bond, 2023). More efforts can be expected from South Africa's Western trading partners to penalise high-emissions export sites with either carbon-intensity or distance-based taxation.

South Africa's National Treasury recognised this in an August 2021 briefing to parliament, in which several sectors were identified as especially vulnerable: iron and steel, cement, fertiliser, aluminium and electricity. The National Treasury announced that it would mitigate the damage by undermining CBAM and "prioritise support to these hard-to-abate sectors"—i.e. increasing state subsidies to high-carbon corporate exporters to offset the damage done in what is allegedly a blatant tax dodge—as well as by imposing its own CBAM on imports (Momoniat, 2021). South Africa's extremely low carbon tax ($0.30/ton in mid-2023, due to stay in place until 2026, compared to the EU's carbon market price of €85–115 during the first half of 2023) will invite substantial import tariffs from trade partners.

5.4.3. Extractivism accounting

It would be relatively easy, in theory, to insist on a broader analysis within fiscal policy to calculate the *full* environmental costs and benefits associated with "natural capital" (e.g. the negative effects of pollution and depletion from these mega-projects) (Gaborone Declaration, 2012). Such calculation can make a government aware of the ecological impacts and address its projects' adverse impacts, especially when the "rights of future generations" are factored in. To illustrate using ecological economic reasoning, if we consider the South African state from the standpoint of a full-cost-accounted, public-sector balance sheet, it is extremely wealthy. A 2018 Fiscal Monitor survey of the wealth of many leading states by the IMF (2018) determined that once a calculation is made not only of net financial assets (i.e. budget surpluses/deficits and public pension funds, minus public debts including state pension liabilities), as is standard in public finance, so too should other non-financial assets be included, such as state-owned natural resources. In that survey, few countries could boast a higher "non-financial asset" wealth ratio than South Africa's 240% of GDP, within state ownership (although many other oil- or mineral-based economies were not calculated). In contrast, South Africa's coloniser Great Britain shouldered a *negative* 120% net worth relative to GDP in 2016.

But the systematic extraction of those non-renewable resources, including coal, imposes a vast cost on South African wealth accounts. The Gross National Income (GNI) captures the national (domestically produced) output of goods and services in a given year. But to sell such "goods"—for instance, South Africa's four leading mineral exports, which are coal (25% in 2020), platinum group metals (22%), gold (17%) and iron ore (11%)—requires extraction, smelting, refining and shipment of non-renewable minerals (ores) and metals, as well as disposal of waste residue. These activities are extremely carbon-intensive, so the burning of coal to facilitate the extraction, smelting and processing of mineral wealth should be factored in by anyone engaged in cost-benefit analyses of mining. Recall that the major consumer of Eskom's electricity is the EIUG (see 5.3.2), comprising 27 firms dominated by mining and smelting operations, typically responsible for around 40% of energy consumption. The largest consumer, BHP Billiton's South32 aluminium smelter, typically uses 5% of the grid for its smelting of imported bauxite. Another factor is the cyclical dynamic of extraction, which is evident in the World Bank's calculation of nearly five decades of mineral and fossil fuel wealth loss in South Africa. In 2011, (non-coal) minerals depleted by $11.7 billion, up from a previous high of $7.2 billion in 2008. That year, coal depletion was measured at $12.3 billion, and in 2011 it was $9.2 billion. So in 2008, these combined resource outflows were $19.5 billion and in 2011, $20.9 billion (World Bank, 2019). These are vast wealth losses, although as a share of GNI South Africa's depletion is less than African economies focused on primary production (e.g. Mauritania, Togo, DRC and Zambia) and especially those with oil sectors (e.g. Angola, the Republic of the Congo and Equatorial Guinea).

What can we learn from perusing these accounts, in which GNI is adjusted to incorporate natural capital? The five major categories in which annual output should be adjusted downwards are: (1) consumption of fixed capital in the form of wear-and-tear depreciation (14.3% of South Africa's GNI), (2) CO_2 damage (4.6%), (3) mineral depletion (1.1%), (4) energy depletion (0.7%), and (5) air pollution (0.4%). Moreover, while energy production is 93% reliant upon coal, which is the main cause of the CO_2 damage, the "benefit" from extracting coal is far less, and has traditionally accrued to multinational corporations like Anglo Coal, BHP Billiton and Glencore. Again, this is now seen as a vulnerability—a stranded asset—so these firms are increasingly selling their coal mines to local black entrepreneurs given that carbon-divestment pressures are rising in their respective headquarter cities of London, Melbourne and even Baar, Switzerland. For instance, in June 2021, BHP Billiton's South32 (2021) spin-off finalised the divestment of its South African coal assets to Seriti Resources. The irony when it comes to coal wealth depletion is that there is negative value to future generations, not only because of the harsh impacts of climate change, but also in the loss of fossil fuel wealth. Although that phrase may appear to be a contradiction in terms, future generations may well possess technology that allows them to extract a component of coal—hydrocarbons—critical for lubricants, pharmaceutical products, synthetic materials and plastics, *so long as the hydrocarbons are not combusted* to release CO_2 and methane. But the demise of the high-BTU coal deposits is another way future generations are harmed by the extraction of non-renewable wealth.

In sum, it is important to assess whether use of the GDP indicator utterly distorts changes in South Africa's wealth beyond recognition, since the carbon-intensive extraction and smelting of minerals contribute such profound *negative* forces associated with non-renewable resource depletion plus pollution, including greenhouse gas emissions. The result of using GDP in a context like this is, as SAFTU (2020b) argued,

> very misleading, and we appeal to [the National Treasury's statistical agency] StatsSA to put a very large asterisk beside it. GDP only counts the output of non-renewable minerals as a positive in this category; it does not count the depletion of the same minerals as a negative. If you count mineral depletion (since it has gone away forever), then the net benefit of mining to the country's actual wealth is negative. Another flaw in GDP calculations is that it ignores the unpaid care work provided by mostly poor and working-class women.... [N]or are pollution costs considered within GDP, so the companies that "externalise" their costs by wrecking the environment do so with applause even though the costs are severe to our and future generations.

This is a seemingly radical conclusion, yet the Gaborone Declaration (2012) (and its Communiqué on Natural Capital Accounting) legitimates precisely this sort of recalculation. Remarkably, it was signed by South Africa's then environment minister, Edna Molewa, along with heads of state or environment ministers from Botswana, Gabon, Ghana, Kenya, Liberia, Mozambique, Namibia, Rwanda and Tanzania. The Gaborone Declaration concedes that since GDP has "limitations of GDP as a measure of well-being and sustainable growth," natural capital should from now on be included in "national accounting and corporate planning." This declaration, in turn, gives climate activists an opportunity to insist that the National Treasury (including its subsidiary statistical agency StatsSA) and the Reserve Bank consider the impacts of environmental factors—including the -4.6% of GNI caused by CO_2 emissions each year, as—very conservatively estimated by the World Bank (2017)—to be not only relevant, but *decisive* in reformulating both climate and macroeconomic policies.

5.4.4. GDP fetishisation: An example

To illustrate the difficulty of reversing the GDP fetishisation mindset, consider how in 2020–2021, Limpopo provincial officials arranged for a preliminary environmental impact assessment (EIA) on the Musina-Makhado Special Economic Zone (MMSEZ). The original proposal called for R400 billion in investment, including a 4,600 MW coal-fired power plant (later reduced in April 2021 to 1,320 MW) designed *not* to feed the national grid but instead to *only supply a variety of smelters*. What was produced in the EIA was incomplete for its lack of attention to both greenhouse gas emissions and the anticipated depletion of natural resources around the MMSEZ site. The author of the EIA openly acknowledges: "The estimation of greenhouse gas emissions was not included in the scope of work. Reference is made to GHG emission reporting regulations as proposed facilities are required to report emissions on the National Atmospheric Emission Inventory System" (EnviroXcellence, 2021, p. 865).

This is a stunning omission; still, to their credit, the Musina-Makhado EIA author acknowledges how damaging the climate implications are (EnviroXcellence, 2021, p. 445):

> The emission over the lifetime of the project will consume as much as 10% of the country's carbon budget. The impact on the emission inventory of the country is therefore HIGH. The project cannot be implemented in the current regulatory confines when considering following:
> - The Nationally Determined Contribution in terms of South Africa's commitment in terms of the Paris Agreement;
> - The Peak Plateau Decline emission trajectory;

- The Integrated Resource Plan, which sets out the planned electricity production capacity of the country;
- When considered on an international level, the project could reduce emissions by as much as 10 million tonnes CO_2e per year, if the plants are built to the recommended emissions intensity specifications.

On the latter point, the EIA author then claims:

> The construction of a coal fired thermal power plant should not be approved unless the plant is fitted with a carbon capture and storage unit that can sequester all emission from the combustion of coal from the starting date of operation. (EnviroXcellence, 2021, p. 480)

Yet there is no operative example anywhere in the world of such CCS operations being technically or financially feasible, especially if there are many tens of millions of tonnes per year of CO_2 to store. In the MMSEZ EIA statement, there was no attempt to assess (a) Limpopo's geological conditions for storing; (b) the implications of the additional energy required (usually 10% in such models) to scrub and then pump CO_2 to the storage site; (c) the risks of storing CO_2 concentrations; and (d) the implications of the anticipated declines in the South African natural wealth accounts—as vast stores of minerals were anticipated for smelting at MMSEZ—included in the EIA (this is not unusual, of course). The Gaborone Declaration should certainly be applied by advocates of climate and environmental justice to rectify this situation.

This is one example of a situation in which attraction of FDI overwhelms environmental-economic common sense in a microeconomic context. One critical task in moving towards a just transition and LFFU is to change the basic economic metrics—i.e. what is being measured when society contemplates what is a successful economy—because economists' reliance upon climate-blind GDP and other orthodox macroeconomic statistics (especially those emanating from StatsSA, the SARB, the National Treasury and financial institutions) represents a profound barrier to change (Fioramenti, 2013).

5.4.5. Climate debt and the social cost of carbon

In addition, there is South Africa's climate debt, both to its own citizens and to the world's (and Africa's) climate victims more generally, a matter which can easily become a fiscal and macroeconomic challenge if either a good government comes to power determined to make good on past injustices (for which all manner of apartheid and colonial reparations claims would be included), or legal liability begins to affect South Africa (like in other jurisdictions). How much is South Africa's climate debt? The level of GHGs that should be considered a "national" liability is very high, especially when adding to CO_2 at least four other major GHGs : CH_4, N_2O, HFCs and PFCs. In 2017, the DFFE (2021) put total GHG emissions that year at 556 Mt, divided as follows: 85% from CO_2, 9% from methane, 5% from nitrous oxide and 1% from fluorinated gasses. If the polluter-pays logic is to apply, there should logically be climate "liabilities" from GHG emissions, determined by a penalty known as the social cost of carbon (SCC). The concept is not uncontroversial but has for at least 15 years become the central variable in assessing notional climate debt. How high to set the SCC, and especially how to assess future damage—using a "discount rate"—are controversial. When the Nobel Economics Prize was given to William Nordhaus in 2018, his own conservative discounting of future life allowed for a 4-degree temperature rise (Hickel, 2018).

Assuming that on average during the 2010s, South Africa's annual emissions were around 500 million tonnes (or 0.5 gigatons), we can make rough estimates of climate debt, i.e. "polluter pay" liabilities owed by those who benefited from emissions in this very skewed economy, and owed to those who are victims of climate loss and damage (in South Africa, Africa and the world as a whole). The costs are typically defined as marginal social damage from emitting one metric tonne of CO_2e. The 2021 Biden Administration's calculation of SCC was $51/metric ton, but in late 2022 the Environmental Protection Agency proposed raising that to $190/ton (Asdourian & Wessel, 2023). However, Kikstra et al. (2021) calculated the SCC to potentially oscillate to as high as $3000 per tonne CO_2 emitted. For an average of 500 Mt CO_2e of annual emissions during the 2000s, the SCC climate debt would be $250 billion, or R3.75 trillion in current terms (using the $500/ton metric) or an even greater $1.5 trillion (R22.5 trillion) according to the more aggressive $3000/ton SCC. In sum, the climate debt being accumulated each year would be between 67 and 402% of national output as measured by GDP, far higher than World Bank (2017) estimates of the cost of South Africa's CO_2 emissions at only 4.6%. The mechanism by which such a massive climate debt is paid (how, to whom, by whom) is still very much subject for debate.

The modalities of raising the funds to cover South Africa's portion of this climate debt, for example through a wealth tax or carbon taxing mechanism, could serve as a useful means of focusing attention. We favour an urgent, stringent *extraction tax* imposed on the dominant domestic and foreign coal, oil and gas companies, as well as an *export tax* imposed on all exported coal and forthcoming exported oil and gas (Sinn, 2012), in addition to the more orthodox *carbon emissions tax*. We favour the inclusion of Scope 3 (downstream) emissions when calculating fossil fuel producers' overall impact; these are currently disclosed by some leading South African firms listed on the Johannesburg Stock Exchange, but are not yet incorporated in environmental impact assessments. It is the very basic "polluter pays" ideology of "internalising externalities" at the basis of microeconomic relations that we should be reminded of, when considering the macroeconomic implications of the climate crisis, and it is here that opportunities arise for not only local LFFU advocacy, but also regional campaigning so that it is clear how South Africa should be acknowledging and paying climate debt owed to neighbours.

5.5. Conclusion

This chapter has pursued a twofold objective: to contextualise South Africa's national context vis-à-vis climate policy and the related key actors (see 5.3); and to explore the key arguments and approaches to LFFU that may prove effective for forthcoming climate action, including innovations (see 5.4). All in all, environmentalists and their allies in the research and advocacy community, working closely with affected trade unionists (or unorganised workers) and communities, have an excellent opportunity to use trade, investment, finance and climate-debt penalties and incentives to broaden the terrain of LFFU campaigning in South Africa and beyond, and to promote inclusive LFFU approaches (see Figure 5.2). Adopting any one of these points of leverage could introduce a series of progressive and radical interventions to challenge the incumbent South African minerals–energy complex and the ruling party.

While the labour movement has few genuine red–green tendencies, SAFTU has articulated strong arguments for compensating workers and communities during a thorough-going decarbonisation process, in alliance with Climate Justice Coalition partners (see Chapter 6). Such movements are critical to destabilise the core regime alliances, but without a systemic strategy and indeed a more far-reaching ideology opposed not only to fossil fuels but also to the system of economic profit they serve, progressive interventions like these may prove insufficient. At some stage, in a society of South Africa's extreme complexity and divergent interests, a broader set of political pressures—say, in the form of a mitigation alliance (MA) (see 8.5)—may yield more promising results for South African climate justice including a genuine just transition from fossil fuels.

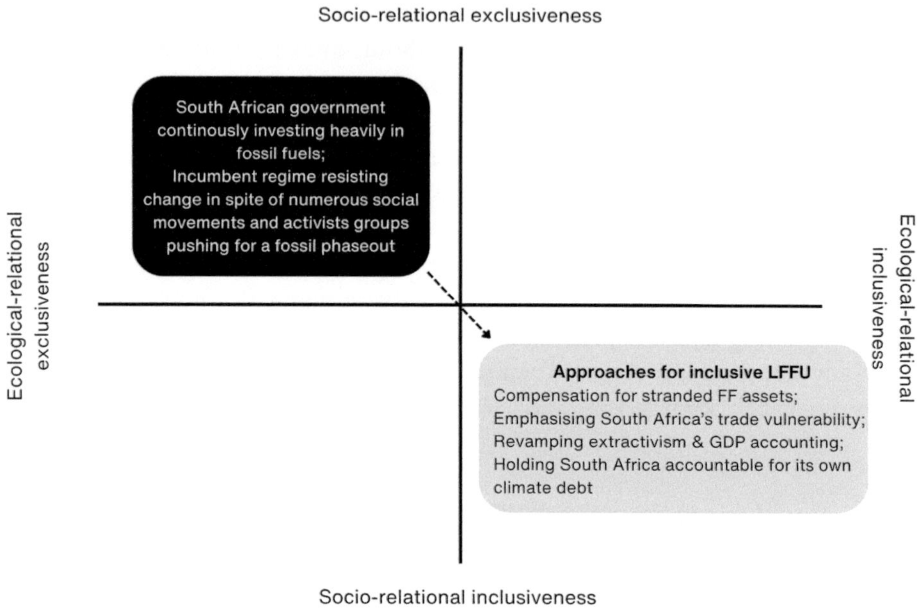

Figure 5.2. South Africa, policies, arguments and inclusive transitions.

References

350 Africa. (N.d.). Climate Justice Coalition. https://350africa.org/the-climate-justice-coalition/

AIDC [Alternative Information and Development Centre]. (2017). One million climate jobs. https://aidc.org.za/programmes/million-climate-jobs-campaign/

Asdourian, E., & Wessel, D. (2023, March 14). What is the social cost of carbon? Brookings Institute. https://www.brookings.edu/2023/03/14/what-is-the-social-cost-of-carbon/

Baker, L., Newell, P., & Phillips, J. (2014). The political economy of energy transitions: The case of South Africa. *New Political Economy, 19*(6), 791–818. https://doi.org/10.1080/13563467.2013.849674

Bega, S. (2019, May 28). SA withdraws doubling of SO_2. *IOL News*. https://old.groundwork.org.za/medarchives/2019/SA_withdraws_doubling_of_SO2.pdf

Bond, P. (2018). Ecological-economic narratives for resisting extractive industries in Africa. In P. Cooney & W. S. Freslon (Eds.), *Environmental Impacts of Transnational Corporations in the Global South* (pp. 73–110). Emerald Publishing. https://doi.org/10.1108/S0161-7230201833

Bond, P. (2022). South Africa's pro-military lobby risks worsening multiple injustices in northern Mozambique. *The Thinker*, 89, 69–80. https://journals.uj.ac.za/index.php/The_Thinker/article/download/1175/754/3442

Bond, P. (2023). South Africa's green climate financing pilots: Flaws in the Just Energy Transition Partnership and Carbon Border Adjustment Mechanism. In J. Jäger and E. Dziwok (Eds.), *Understanding green finance*. Edward Elgar.

Climate Action Tracker. (2022). South Africa. https://climateactiontracker.org/countries/south-africa/

CJC [Climate Justice Coalition]. (2021). *A green new Eskom: Transforming power, transforming South Africa*. https://350africa.org/files/2021/05/GreenEskom_Booklet_new2.pdf

Cock, J. (2019). Resistance to coal inequalities and the possibilities of a just transition in South Africa. *Development Southern Africa, 36*(6), 860–873. https://doi.org/10.1080/0376835X.2019.1660859

Cronje, J. (2022, March 2). Coal feet. *News24*. https://www.news24.com/fin24/companies/coal-feet-limpopo-industrial-park-backed-by-china-ditches-plans-to-build-power-station-20220302

De Ruyter, A. (2023). *Truth to power*. Penguin.

DFFE [Department of Forestry, Fisheries and the Environment]. (2021). National GHG inventory report, South Africa, 2017. https://www.dffe.gov.za/sites/default/files/docs/nir-2017-report.pdf

DMRE [Department of Mineral Resources and Energy]. (2019). Integrated Resource Plan. https://www.energy.gov.za/irp/2019/IRP-2019.pdf

Editor BizNews. (2023, February 23). Giving unto Caesar—ANC donors include Naspers, Harmony Gold et al. *BizNews*. https://www.biznews.com/leadership/2022/02/23/naspers-anc-donations

Ensor, L. (2023, June 2). Pravin Gordhan still hopeful despite failure to resolve dispute over trains in China. *Business Day*. https://www.businesslive.co.za/bd/national/2023-06-02-pravin-gordhan-still-hopeful-despite-failure-to-resolve-dispute-over-trains-in-china/

EnviroXcellence. (2021). Musina-Makhado Special Economic Zone (MMSEZ). Environmental impacts assessment report. Final report, version 03. LIM/EIA/0000793/2019. http://www.enviroxcellence.co.za/assets/document/FINAL%20EIAR%20EXS%201%20(1).pdf

Evans, S. (2021, October 5). Which countries are historically responsible for climate change? *Carbon Brief*. https://www.carbonbrief.org/analysis-which-countries-are-historically-responsible-for-climate-change/

Fine, B., & Rustomjee, Z. (2018). *The political economy of South Africa: From minerals–energy complex to industrialisation*. Taylor & Francis.

Fioramonti, L. (2013). *Gross domestic problem: The politics behind the world's most powerful number*.

Gaborone Declaration. (2012). Summit for Sustainability in Africa. https://www.cbd.int/doc/champions/gaborone-declaration-botswana-en.pdf

Govender, P. (2013). Eskom overview and National Development Plan SIPs. Presentation to the 2013 Construction Engineers South Africa Annual Conference. Johannesburg. https://www.cesa.co.za/sites/default/files/20131112_Poobie%20Govender%20-%20CESA%20Presentation_FINAL.pdf

Haffajee, F. (2023, March 6). Ramaphosa gets second bloody Phala Phala nose in two weeks. *Daily Maverick*. https://www.dailymaverick.co.za/article/2023-03-06-ramaphosa-gets-second-bloody-phala-phala-nose-in-two-weeks/

Hickel, J. (2018, December 6). The Nobel Prize for climate catastrophe. *Foreign Policy*. https://foreignpolicy.com/2018/12/06/the-nobel-prize-for-climate-catastrophe/

IMF [International Monetary Fund]. (2018, October 4). Fiscal Monitor: Managing public wealth. https://www.imf.org/en/Publications/FM/Issues/2018/10/04/fiscal-monitor-october-2018.

Kikstra, J. S., Waidelich, P., Rising, J., Yumashev, D., Hope, C., & Brierley, C. M. (2021). The social cost of carbon dioxide under climate-economy feedbacks and temperature variability. *Environmental Research Letters, 16*(9), 094037.

Leshoro, D. (2023, May 15). Climate Commission seems to back extending life of coal power plants amid load shedding crisis. *News24*. https://www.news24.com/citypress/business/presidential-climate-commission-backs-extending-life-of-coal-power-plants-amid-load-shedding-crisis-20230515

Momoniat, I. (2021, August 18). Carbon tax policy update and sustainable finance initiatives. Portfolio Committee On Environment, Forestry And Fisheries, National Treasury, Government of South Africa. https://static.pmg.org.za/210818_-_PC_EFF_-_NT_-_FINAL_Climate_Change_and_Carbon_Tax_Presentation_to_Parliament_NT_170821.pdf

Parry, I., Black, S., & Vernon, N. (2021). Still not getting energy prices right. Working Paper 2021/236, International Monetary Fund.

Pijoos, A. (2022, June 29). Amid load shedding pain, more than 50% of Soweto residents are illegally connected to grid—Eskom. *News24*. https://www.news24.com/news24/southafrica/news/amid-load-shedding-pain-more-than-50-of-soweto-residents-are-illegally-connected-to-grid-eskom-20220629

Rafey, W., & Sovacool, B. K. (2011). Competing discourses of energy development: The implications of the Medupi coal-fired power plant in South Africa. *Global Environmental Change, 21*(3), 1141–1151.

Rempel, A. (2023). *Leaving fossil fuels underground in South Africa: From a climate debt to an unsettled stranded asset debt*. PhD diss., University of Amsterdam. https://hdl.handle.net/11245.1/626e6e22-cebe-4f3f-a8ae-ea13e753b9f0

Republic of South Africa. (2021, September). South Africa first nationally determined contribution under the Paris Agreement. Updated September 2021. https://unfccc.int/sites/default/files/NDC/2022-06/South%20Africa%20updated%20first%20NDC%20September%202021.pdf

SAFTU [South African Federation of Trade Unions]. (2020b, December 20). SAFTU Statement on StatsSA's latest misleading GDP figures. https://saftu.org.za/saftu-statement-on-statssas-latest-misleading-gdp-figures/

SA News. (2018, February 7). President lauds Total's discovery of gas condensate. https://www.sanews.gov.za/south-africa/president-lauds-totals-discovery-gas-condensate

SA News. (2021, July 5). Mining returns must promote more employment-friendly activities. https://www.sanews.gov.za/south-africa/mining-returns-must-promote-more-employment-friendly-activities

Sinn, H. W. (2012). *The green paradox: A supply-side approach to global warming.* MIT Press.

Sinwell, L., Nwane, T., & Maggott, T. (2022). Electing to electrify: Unpacking the local crisis and state response in Sun Valley, Soweto. *Politikon, 49*(4), 337–349. https://doi.org/10.1080/02589346.2022.2149114

Smit, S. (2019, August 23). Mischief managed: Cyril's take on CR17 fund flows. *Mail & Guardian.* https://mg.co.za/article/2019-08-23-00-mischief-managed-cyrils-take-on-cr17-fund-flows/

South32. (2021). South Africa energy coal divestment update. https://www.south32.net/docs/default-source/annual-reporting-suite/south-africa-energy-coal-divestment-updatecdfadf58e3004e5e9fb168d21a38b35d.pdf?sfvrsn=28fdd000_1

Swilling, M., Musango, J., & Wakeford, J. (2016). Developmental states and sustainability transitions: Prospects of a just transition in South Africa. *Journal of Environmental Policy & Planning, 18*(5), 650–672. https://doi.org/10.1080/1523908X.2015.1107716

Van Tilberg, L. (2023, May 23). Karpowership deal could cost South Africa R200 billion and do little to solve loadshedding. *BizNews.* https://www.biznews.com/interviews/2023/05/23/karpowership-deal-cost-r200-billion-mileham

Winkler, H., & Marquard, A. (2007). Energy development and climate change: Decarbonising growth in South Africa. UNDP Occasional Paper, 2007/40. https://hdr.undp.org/system/files/documents/winklerharaldandmarquardandrew.pdf

Winkler, H., & Marquard, A. (2020). Climate finance to transform energy infrastructure as part of a just transition in South Africa. University of Cape Town. https://open.uct.ac.za/handle/11427/32176

World Bank (2017). *The little green databook 2017.* https://openknowledge.worldbank.org/handle/10986/27466

World Bank (2019). Adjusted savings: Mineral depletion (current US$). https://data.worldbank.org/indicator/NY.ADJ.DMIN.CD?locations=ZA

World Bank (2021). *The changing wealth of nations 2018.* https://www.worldbank.org/en/news/feature/2018/01/30/the-changing-wealth-of-nations-2018

York, R., & Bell, S. E. (2019). Energy transitions or additions? Why a transition from fossil fuels requires more than the growth of renewable energy. *Energy Research & Social Science, 51,* 40–43.

6. South Africa and Social Mobilisation for LFFU

Malin Olofsson

Abstract

This chapter demonstrates how since the 2010s there has been an expansion and convergence in social mobilization for LFFU across South Africa, including labour movements, feminist movements and environmental and social movements. This is particularly impressive, considering the deep and widespread social and economic deprivations the country faces. While the arguments, approaches and strategies employed by the different actors are diverse, many have managed to unite in broad coalitions for climate justice. Beyond the work done around fighting single-issue cases, through the successful partnerships between community-based, national and international organisations, significant arguments engage with the politics of the possible and shift beyond a just transition to transformative ideas for a post-capitalist world focused on eco-socialist and feminist alternatives. This chapter illuminates important developments for advancing the energy transition in a just and inclusive manner in South Africa, but also some controversies and challenges that prevent the movement from gaining more traction.

Keywords: fossil fuels, resistance, social movements, politics of ideas, South Africa

6.1. Introduction

The devastating impacts of climate change together with expanding extraction of fossil fuels have led to the rapid expansion of environmental and climate justice movements across Africa since 2000 (see

---

Gupta, J., Hogenboom, B., Rempel, A. & Olofsson, M. (eds), *Leaving Fossil Fuels Underground: Actors, Arguments and Approaches in the Global South and Global North.*
Taylor & Francis Group, 2024
doi: 10.5117/9789048560370_CH06

3.3). What started out as siloed initiatives in response to site-specific environmental harms has grown into an ever more networked movement within Africa. The founding of the Pan African Climate Justice Alliance (PACJA) in 2008 is a testament to this convergence of movements. It brought together over a hundred civil society organisations to influence regional and global climate policy. PACJA has been a progressive force pushing forward climate justice in Africa and has been outspoken in UN climate negotiations since Copenhagen in 2009, calling out the inadequacy of the commitment that came out of Paris and subsequent UN climate COPs (Mwenda & Bond, 2020). The slow progress of global climate negotiations in tackling the climate emergency has only invigorated the movement. Climate action "paralysis from above" has been met with a countermovement from below where new and exciting forms of climate justice activism have emerged, much of which is to be found in Africa (Mwenda & Bond, 2020, p. 3).

South Africa is an exceptional case due to its massive and disproportionate carbon footprint compared to the rest of the continent. It is Africa's highest greenhouse gas emitter and the 14th highest globally, relying heavily on coal. As discussed in Chapter 5, the minerals–energy complex was and is the purview of a small elite of national and international actors, heavily supported by the state and particularly the pro-fossil fuel stance of the Department of Mineral Resources and Energy (DMRE). In 2020 the South African government provided R172 billion (that is 172 billion South African Rand or ZAR) in energy subsidies, most of which went directly or indirectly to fossil fuels. The total fossil fuel revenues in 2019 amounted to R100.5 billion or 7.4% of general revenue. The social costs of fossil fuel use are five times higher than the revenues, with an estimated cost to society of R550 billion (Bridle et al., 2022, p. vi). These social costs mostly refer to the impacts of climate change-related severe weather events and the associated local air pollution effects from fossil fuel combustion. These are disproportionately experienced by the poor and marginalised, which in South Africa refers to the black majority.

This carbon-intensive economy and its severe social costs need to be situated within a context that has the highest global inequality levels with 63% of the people living below the poverty line (Budlender et al., 2015). South Africa also faces a devastating energy crisis since 2007. Rolling backouts or "load shedding" currently last for up to 12 hours a day as the national electricity utility—Eskom—cannot meet demand. Electricity prices have risen 653% from 2007 to 2022, more than five times inflation (Moolman, 2022), affecting the affordability of energy for the majority and plunging many into darkness. This context has created one of the most difficult places to advocate for climate justice (Mwenda & Bond, 2020, p. 7). Climate issues are often side-lined as poor South Africans take to the street to demonstrate their discontent with the prevailing socio-economic conditions, the persistent service-delivery problems and growing political contestation. Thirteen categories of protest action have been identified, the highest being about labour, crime/policing and municipal services (Lancaster, 2018), while climate and environmental issues did not even feature.

The South African state occupies a contradictory position. The Presidency is advocating for a progressive-sounding Just Energy Transition Investment Plan (JET IP), which aims to raise R1.5 trillion for the initial period from 2023 to 2027 to enable the country to shift toward renewable energy, green hydrogen and electric vehicle production (The Presidency, 2022). While this plan appears to show that South Africa is serious about moving towards climate actions and LFFU, the minister of mineral resources and energy simultaneously put out a new call for proposals for gas extraction (Faber, 2022), as part of his ongoing push for the extraction of fossil fuels as a "game changer" for the South Africa economy.

This wider socio-economic and political context in South Africa has accelerated in many instances the convergence of diverse climate movements in society. Increasingly labour movements, feminist movements and environmental and social movements are making the links to climate change and LFFU, albeit that for some this is more explicit and central than for others. Many of these movements were established since 2015 in response to the growing convergence of crises. This chapter dives into this "movement from below" (Bond, 2012) by exploring the actors involved in the climate justice movement, the arguments they advance and the approaches they employ. It by no means provides an exhaustive analysis of the climate justice movement in South Africa, but attempts through a selective sample of movement actors to gain insights into the dynamics on the ground. In this way it illuminates some of the exciting and promising developments for advancing the energy transition in a just and inclusive manner, but also some of the controversies and challenges that prevent the movement from gaining more traction and driving regime- and landscape-level transitions (see 2.2.3). The chapter first briefly presents the links to the analytical framework set out in Chapter 2 (see 6.2), and then it describes the contours of the climate justice movement in South Africa (see 6.3), maps out the actors involved and related arguments and approaches to LFFU (see 6.4) and draws conclusions (see 6.5).

6.2. Building on the analytical framework and methods

This chapter applies an inclusive development framework (see 2.2.2), with a particular focus on the arguments and approaches of environmental and climate justice movements in South Africa, in the context of the energy transition. Through unpacking the different arguments and approaches of the movement, I acknowledge the internally differentiated nature of the movement (Cock, 2019). To this end, different class, identity and ideological interests are central, which necessarily complicate any simple notion of a unified climate justice movement in South Africa. These dimensions guide and sharpen the analysis of the social, ecological and relational inclusiveness of the inclusive development framework. The framework invites an enquiry into the social, ecological and relational dimensions of development. Social inclusiveness in the context of the energy transition refers in particular to workers and communities living in the vicinity of coal mines and related industries, but also to poor and marginalised citizens more generally and how they stand to be affected by an energy transition. Ecological inclusiveness relates to how people, mostly those living in the vicinity of the coal mines, are affected by the pollution of natural resources and the loss of access to key natural resources such as land and water. And, lastly, relational inclusiveness here is connected to the way and extent to which social movements and other non-government actors fighting for environmental and climate justice are able to influence and shape the state and other national and international actors' positions and actions on energy and climate policy. These three dimensions are used to explore how representatives from the different movements *articulate* their struggles, what *strategies and tactics* they employ and what *traction* these have across spatial-geographical scales and governance levels.

In analysing the different arguments and approaches to LFFU employed by different actors, I use a scalar analytic lens to look at international, national and local actors and their related argument and approaches to LFFU. This potentially foregrounds important differences and sheds light on the controversies and challenges that exist between different movement actors across scales. Fieldwork for this study took place in July–August 2022. This (limited) time presented the opportunity to get some broad stoke impressions of the dynamics in the climate justice movement in South Africa. Semi-structured interviews were conducted with 11 key informants from community-based organisations (CBOs), local NGOs, INGOs, activists and a labour union federation. In addition, participant observation, media coverage and a document review were used as data sources.

6.3. The contours of the climate justice movement in South Africa

It was a cold wintery morning in Pretoria on 18 July 2022. This day, also known as Mandela Day, is widely celebrated as marking the legacy of the man who steered the country towards democracy. I set out to join a march to the Union Building—the office of the Presidency—coordinated by the Climate Justice Coalition (CJC). The CJC is relatively new, established in 2020 through the initiative of various organisations following the Global Climate Strike in Gauteng in September 2019. It is a broad coalition of 40 organisations that include trade unions as well as grassroots, community and non-profit organisations. Its mission is to build and strengthen the climate justice movement, and it does this by bringing the various diverse movements together to jointly push for radical system change to ensure climate justice. The diversity of tactics employed across the different member organisations (including litigation, advocacy, education, training, mobilising, organising, campaigning and non-violent direct actions) is seen as one of the strengths of the CJC (CJC, 2022). The CJC has identified two main campaigns—"Green New Eskom" and "Uproot the DMRE"—as being best positioned to advance the cause of climate justice and to deepen and strengthen the broader movements. The march on this particular morning was around the latter.

"Uproot the DMRE" places the minister of the DMRE, Gwede Mantashe, at the centre of the current energy crisis in South Africa. Under his leadership the DMRE is seen as blocking a just transition to a renewable energy future by locking the country into a fossil fuel future, through its "expensive, polluting and outdated energy future" (Uproot the DMRE, n.d.). The march was organised to deliver a petition to the Presidency with a list of demands, including the implementation on an emergency renewable energy plan to end load shedding and to replace the minister himself. I was not greeted by swathes of people, but rather a small gathering of mostly NGO and INGO folk. Various community organisations from around Gauteng and labour movements joined the procession as we made our way through the streets of Pretoria to the union building. As we gathered just outside of the fence that surrounds the Union buildings, representatives from the different organisations took the microphone. The representatives speaking on behalf of the Voices of the Poor and Concerned Residents, The Ivory Park Residents Committee and the Soweto Electricity Crisis Committee raised similar shared concerns around the high price of living, worsening blackouts or load shedding, electricity tariff increases and deepening poverty. As one speaker noted:

> Mandela Day is supposed to be a day that represents freedom, but the condition of people today is not one of freedom. We thought by now we would be having a better life but we don't.... Apartheid was evil but this democracy is also evil.

The same themes were reiterated by the different speakers representing the community organisations, that of the growing struggles of the poor and working classes to sustain their basic social reproductive needs within the context of high inflation, joblessness and pervasive poverty. Other speakers linked the everyday struggles of the poor and working classes with the wider systems of oppression, climate and justice issues. The speaker of the South African Federation of Trade Unions (SAFTU) focused on capitalism as the systemic cause of the current crisis. "We are passing through one of the biggest crisis of all time, a crisis of poverty, inequality and climate, ultimately a crisis of capitalism." He highlighted the class dimension of the energy crisis, whereby electricity is shut off from poor working-class neighbourhoods in Soweto as a punitive measure during load reduction, referred to elsewhere as "energy racism" (Maggott et al., 2022). "The capitalist class wants to offload the burden of this crisis on the working classes." Moving from a critique of capitalism and the accompanying system of oppression, the co-founder of the Debt for Climate movement shifted the focus to the global financial flows and South Africa's debt burden. "We in South Africa pay 200 billion on interest payments to the World Bank and the International Monetary Fund.... [W]e need to cancel the evil, immoral and odious debt. Cancelling the debt could finance the just transition and pay a basic income grant." In this way the movement is pushing for climate action through a reallocation of debt repayment towards addressing the climate crisis.

A final speaker from the local chapter of 350 Africa eloquently linked climate change to the water crisis experience in Cape Town (2015–2018) and currently in the Eastern Cape, Grahamstown and Gqeberha, to the recent floods in Durban, and to the current coal-generated energy system that is failing to provide reliable, sufficient or affordable electricity. His message was more solution oriented, highlighting how renewable energy can be bought onto the grid relatively quickly and affordably and how a just energy transition will also be able to provide jobs. The DMRE with Minister Mantashe at its helm was singled out as the major stumbling block. The secretary of the CJC took this point further, calling the DMRE a criminal organisation for preventing the rapid and just transition to renewable energy. Finally, the petition, which demands the president implement an emergency renewable energy plan to end load shedding, replace the current minister of energy and overhaul the DMRE and rapidly move towards a more socially owned renewable energy-powered economy, was handed over to an official flanked by security police.

The diversity of actors present at this march points to the diversity of actors and organisations in South Africa whose struggles are linked in varying degrees to LFFU. These links are more or less explicit in their respective struggles. The next sections map the different actors from a scalar perspective and then elaborates the overarching arguments of the different movement and how these inform their approaches.

6.4. Actors, arguments and approaches to LFFU

Many organisations work on climate justice-related issues in South Africa. They are increasingly interconnected through joint campaigns, providing support and solidarity to one another when it comes to overlapping interests. However, there are inevitable tensions between movements that pose a challenge to movement building. International organisations (mostly with headquarters in the Global North), regional, national, and local community-based and labour-based organisations are all active on the ground in South Africa. The scale at which they are positioned influences their mandates, accountability structures and focus, and ultimately their arguments and approaches for LFFU, with very real local political implications for building the movement and ensuring a just energy transition.

6.4.1. International NGOs: Strategically supporting the movement from below

A common strategy across international non-governmental organisations (INGOs) operating in South Africa is linking with local groups/organisations with whom their existing strategy and programming align. Most focus on specific campaigns and projects. Greenpeace has taken an issue-based approach and runs crowdfunding campaigns for issues such as Shell's seismic blasting on the Wild Coast on the eastern side of the country, and a petition to the president to fast-track the shift to renewable energy. Its approach is largely technical, focusing on getting renewables onto the national grid, demanding that all red tape be removed from the bidding process for renewable independent power producers (IPPs) and facilitating the ability of municipalities to purchase renewable energy directly from them. The World Wildlife Fund (WWF) is similarly focused on technical innovations to facilitate the transition to renewables. It convenes high-level dialogues and ambitious target setting by government and business while supporting companies to cut emissions and adapt their investment strategies away from fossil fuels. Oxfam's focus has been more towards movement building amongst communities facing the impacts of climate change. Such support is often in the form of grants to support their struggles and building capacity amongst communities to take legal action against fossil fuels companies when environmental rights stand to be undermined (see also Chapter 3 on court cases by socio-environmental organisations). The victory of coastal communities in the court case against Shell in September 2022 revoked the company's rights to seismic blasting off the Wild Coast and illustrates the importance of such support. These actions can be broadly seen as not necessarily engaging directly with anti-extractives politics but rather engaging with project damage mitigation and community consultation and consent that do not necessarily grapple with the more radical politics of the climate justice movement.

It is clear that 350 Africa is somewhat different in its approach to the other INGOs as it is more a global movement than an organisation in the conventional sense. Its strategy is one of movement building to fighting climate change. However, it goes beyond fighting, as a movement leader stated in an interview:

> There is an emerging understanding that if you are working to stop things your work will never be done. There is a shift in the organisation now towards starting things and looking to solutions and what that might look like. That's exciting work and in South Africa that focuses on the just transition because that is the big solution we need in South Africa.

It positions itself within the "politics of the possible" (Sangari, 2002) and explicitly highlights the justice aspect of the "just energy transition," arguing for such a transition to be based on public ownership of renewables, accompanied by progressive ideas around financing the transition. These include climate debt and reparations both from the Global North but also from South African companies. In the latter case it argues for a wealth tax on companies that have benefitted from highly subsidised electricity costs,[1] such as smelters and mining companies, in addition to cancelling Eskom's debt and using this to finance a just transition.

1 Eskom tariff prices to mining and industry are just over half what they charge residents and agriculture (Maggott et al., 2022, p. 61).

Another example of 350 Africa initiating things is its role in influencing the setting up of the national Climate Justice Coalition (CJC), the largest coalition of climate justice organisations in the country. It has strategically acted as a catalyst, birthing something that could subsequently stand on its own. As a leader of 350 Africa shared, "to build a truly diverse and South Africa coalition, 350 [Africa] needs to be a part of and not the umbrella organization." This stance is also due to the attempts from the government to delegitimise the climate justice movement. The DMRE minister has accused the CJC of being funded by foreign forces intent on destabilising the DMRE and accusing it of personally attacking the minister. The minister went as far as to threaten to sue the secretary of the CJC and the organiser at 350 Africa for spreading false claims. While this at once illustrates the reach and impacts of the climate justice movement that the minister felt so personally threatened by their critique of his current energy policy, it also emphasises the importance of the South African Climate Justice movement being seen as a truly local one, that represents local interests and peoples. This was in part why the CJC became the umbrella organisation with 350 Africa participating as a regular member.

350 Africa has also played a key role in the fair finance coalition, which focuses on the transparency of development finance institutions, in particular that of the African Development Bank, the Development Bank of Southern Africa, the Export Credit Insurance Corporation, the Industrial Development Corporation and the New Development Bank,[2] with the aim to ensure investments are socially and environmentally responsible, with a specific focus on issues of climate change and transparency. 350 Africa is arguably the most politically progressive of the INGOs working in South Africa and has contributed to the movement-building efforts through the CJC and more generally through partnering with local struggles in their efforts to build an inclusive people-powered movement:

> To change everything, we need to be inclusive. We need to talk to people in their languages, in a way that people understand through channels of communication that will reach them.... People power is in the DNA of 350 [Africa] and we all understand that we need to bring more people on board. However, it is a big challenge in a country where there is mass unemployment and poverty, and people are often focused on putting a

2 The Fair Finance coalition is made up of a number of environmental justice grouping that include the Centre for Environmental Rights, the African Climate Reality Project, the Centre for Applied Legal Studies, Oxfam South Africa and Earthlife Africa amongst others.

> meal on the table and finding a job. So, while we want to bring them on board, we also understand that it is not their responsibility—it is the government's responsibility, it is the corporates' responsibility. They should be doing the work (Interview with a leader of 350 Africa).

6.4.2. National NGOs and CBOs: "Nothing about us without us"

The links between national NGOs and CBOs are increasingly blurred as they often work hand in glove as are the links here between environment and climate justice. Many national and local NGOs have galvinised their work around the environmental impacts of the fossil fuel industry on communities affected by coal mining and related environmental damage. The South Durban Community Environmental Alliance (SDCEA), Earthlife Africa, Mining Affected Communities United in Action (MACUA) and WoMin are among those that were explored during this research. A common feature in their approach is to support communities to organise, mobilise and resist. This is enacted in various ways from building solidarity within and between communities across scales, through awareness raising and education and linking CBOs with legal support to fight their specific struggles in the courts. As the director of Earthlife Africa explained, "Our biggest focus is that people are informed and can take up issues themselves.... Our education focuses on linking everyday struggles with the bigger issues. If people are informed, they will mobilise themselves." A similar objective drives the work of MACUA, whose program director explained:

> People don't give a shit about climate change when they are worried about jobs and having food on the table—that's the reality. Mining is currently not benefitting communities. A just transition doesn't mean anything to communities. They don't have access to energy that is generated or benefit from the opportunities in the mining.

The everyday impacts of a changing climate are profoundly experienced by communities. However, as mentioned, linking these impacts to the burning of fossil fuels and coal mining is a key strategy in NGO efforts to support communities in mobilising. As a community organiser explained, when the lack of access to water in a community is linked to the fact that water is being diverted to wash coal in a nearby coal mine, people get outraged and can direct their anger in a specific direction. Making such links is of course critical to movement building in communities, however, a critical next step for local NGOs is to support the partnership between communities and environmental legal firms that can support communities in bringing their cases to the courts. The courts are seen as a key site to fight for environmental and social justice but also importantly there is the hope that by winning a case it may set precedents and redefine policy through litigation. The litigation supported by Earthlife has focused on procedural grounds, namely the lack of adequate public participation in project planning and inadequate environmental impacts assessments.[3]

While NGOs in this space aim to show the links between fossil fuels, climate impacts and the everyday struggles faced by communities across South Africa, most mining-affected communities do not have access to the energy that is generated from the coal mining operations or benefit from the opportunities in the mine. Hence, their struggles are not necessarily anti-coal mining per se, but rather converge around having a stake in mining operations and access to the benefit from mining. Their arguments centre around procedural and distributive processes, and their strategy is to build a mass movement to force change in the sector. A MACUA and Women Affected by Mining United in Action (WAMUA) slogan aptly sums up their position: "Nothing about us without us." This has recently motivated them to also turn attention to the minerals required for the energy transition, particularly for renewable energy technology. They use social audits to determine how communities are affected by mining and in turn how they could have a greater stake in mineral extraction required for the energy transition and in the reskilling of workers.

Important work is also being done by NGOs and community organisations in centring women and how they are disproportionally affected by the impacts of mining and climate change while being systematically excluded from the benefits of mining. This has brought a distinct ecofeminist lens to the struggles for climate justice. Rural women in particular are at the forefront of the impacts of extraction and climate change as they lose access to land and water. WoMin and WAMUA have both centred women in their efforts to fight for climate justice, higlighting that women are affected differently by mining than men, but also that they organise differently. They thereby bring a distinctly gendered lens to the struggle for climate justice.

In the urban context, one of the many struggles poor and working-class people have converged around is energy access. Most notable is the Soweto[4]

3 See the Thabametsi case (Earthlife Africa Johannesburg vs The Ministry of Environmental Affairs, 2017) and the Nuclear case (Earthlife Africa Johannesburg and Southern African Faith Communities vs The Ministry of Energy, 2017).

4 Soweto, a poor, black township on the outskirts of Johannesburg.

Electricity Crisis Committee, which emerged in 2001 to address this issue. It has highlighted the racist and classist practices of Eskom through the implementation of a system of "load reduction." This is a system of semi-planned power outages that target black working-class townships, allegedly for not having paid their electricity debt and for "illegal" connections. This is over and above the national load shedding schedule, which is the ongoing national power outages affecting the entire country. In a report that the Electricity Crisis Committee was involved in drafting, they highlight how the practice of Eskom systematically disadvantage those already disadvantaged, highlighting how "load reduction" is both racist and classist and thereby a discriminatory form of energy saving. They conducted research into the recent (2020–2021) spate of energy cut-offs by Eskom and suggest

> black and working-class communities both pay more for their supply of electricity and are more likely to be cut off for extended periods (up to 11 months or more) due to Eskom's "load reduction" programme and slow response to technical problems (Maggott et al., 2022, p. 6).

The Soweto Electricity Crisis Committee has been important in building a movement amongst local resident that temporarily addresses the issue of lack of electricity access through reconnecting residents to the national grid but also unifying and amplifying the struggles of local residents in protest of the current energy supply practices that systematically disadvantage the already disadvantaged. As an organiser in the Soweto Electricity Crisis Committee emphasised:

> In SA the only language they [government] can listen to is radical actions. Whenever you need to sit with them in boardrooms or meetings, they are very slow to respond. That's when you take it to the street and request radical change. That is when they listen to you.

So their struggles are regularly taken to the streets, be it in protest around inadequate service delivery, lack of transparency in the billing of energy via prepaid electricity meters or the price of energy.

6.4.3. Labour movement: Envisioning an eco-socialist future

The concept of a just transition was initially put forward by the labour movement in the 1970s as a part of the struggle for addressing workers' health and safety in polluting industries in the US. Since then the labour movement has been responsible for putting the concept on the global agenda when it comes to climate change and related mitigation measures (COSATU, 2022). A good example is that the union movement pushed for the just transition principles to be included in the Paris Agreement, which acknowledges the need for "taking into account the imperatives of a just transition of the workforce and the creation of decent work and quality jobs in accordance with nationally defined development priorities" (UNFCCC, 2015). Since 2010 the union movement in South Africa has engaged with the implications of climate change and climate justice more broadly. In 2011 the government launched a white paper outlining its national climate change response (South African Government, 2011) and in the same year the country hosted the COP17 in Durban and released the first Integrated Resource Plan (IRP), which articulated the way forward for the country's energy mix (see Chapter 5). It was around this time that unions, particularly the National Union of Metalworkers of South Africa (NUMSA) but also the Congress of South African Trade Unions (COSATU), realised they were on the back foot with regard to the energy and climate debates taking place nationally and globally. They decided to address the lack of clear policy positions on climate change and the position of the working classes. The labour movement wanted a union approach to COP17, especially with regard to mitigation and adaptation, and set up a research and development group to take this on (Satgar, 2015). It was also the momentum generated by the Alternative Information and Development Centre's 1 Million Climate Jobs campaign that mapped out a pathway for a just transition to a low-carbon economy as a way to combat unemployment and climate change simultaneously that got unions actively involved (Satgar, 2015). As the policy and research officer of the South African Federation of Trade Unions (SAFTU) acknowledges, "I think it [the just transition] has actually been forced on us, and by that I mean forced through the actions of communities but also the actions of the government."

NUMSA was the first union to engage with a just transition and what this would mean for workers. This is largely due to its constituencies, which include Eskom and mine workers, who will be directly impacted by climate and energy policies and therefore have very specific demands in this regard (Interview with SAFTU policy and research coordinator). By 2012, NUMSA adopted two important resolutions to guide its position on climate change and justice: "Climate Change and Class Struggle" commits to finding climate justice solutions from below in order to move to a low-carbon economy; and "Building a Socially Owned Renewable Energy Sector in South Africa" centres the importance of a renewable energy sector being owned and controlled by workers (Satgar, 2015). SAFTU acknowledges that it is only recently—due largely to pressure from its affiliates, in particular, NUMSA—developed policy positions on climate change and the just transition. These are articulated in key grievances linked to a set of demands to government set out in a Section 77 application in 2020. Many of these grievances and the framing of the current demands overlap with COSATU's more elaborated "Just Transition Blueprint for Workers" (COSATU, 2022).

Both NUMSA and COSATU emphasise the need for a *deep* and *transformative* just transition that acknowledges the need for protecting and reskilling workers and communities directly affected by the transition to renewables. However they go beyond a narrow focus on justice for workers and communities by framing justice more broadly, proposing a break from the capitalist mode of production that is driving the climate crisis and rests on the exploitation of workers and nature. COSATU are explicit about a post-capitalist future being one which centres eco-socialism, which restructures the economy to support people and not profits, centres meeting human needs within the planetary boundaries and acknowledges the interconnections between humans and nature. Some of the key principles of a deep and transformative just transition is that it is people driven and centred, involving participatory planning, public finance, social ownership, progressive regulations and taxation. A deep and transformative just transition involves a halt to all current and planned carbon-intensive mega-projects being supported and promoted by the DMRE, and instead a move towards LFFU and 100% renewable energy that is socially owned. But while aspirations of an eco-socialist future are clear in the demands and policy documents, SAFTU acknowledges that this is a challenging position for a union federation whose bread and butter work is ensuring jobs and livelihoods are created and secured:

> We are anti-privatisation and also anti-mining, generally. But we do want more job creation, so it's a very difficult set of policy positions to arrive at because we do require more jobs in the mining sector ... but you also have to have a policy position which we have, which says that energy generation needs to be hybridised so that you have more and more renewable solutions. But you also need coal because you're in South Africa, which is not as industrialised as those countries that are phasing out coal. So, you almost need a balance, saying that corporations that produce coal need to be taxed more and the royalties need to be more. The production of coal needs to be more controlled by government ... and we need it to be more socialised to the South African context so that it's actually giving energy to those who need it. (Interview with a SAFTU representative)

The union movement thus provides an important vision for the way forward with regards to LFFU in South Africa, although there are contradictions and tensions within the movement in the short term as expressed in this quote.

6.4.4. Coalitions: Enabling convergence around a "deep just transition"

The Climate Justice Coalition (CJC) and the Climate Justice Charter Movement (CJCM) are two of the main overarching movement-building organisations that focus on uniting the diverse movements and organisations to achieve a just energy transition and climate justice. They have many overlapping objectives and aspirations; however, they have quite different strategies. The CJC focuses on bringing diverse movements and organisations together (currently 44) and has galvinised around two main campaigns, "Uproot the DMRE" and "Green New Eskom." In line with the work of 350 Africa to move from resisting to creating, both campaigns are focused on tangible steps to move towards a just energy transition, by transforming the parastatal energy utility and getting a progressive leadership at the helm of the Ministry of Mineral Resources and Energy to steer the just transitions. People's power, democracy and public ownership are key ideas that underpin both these campaigns.

CJCM is broad-based in reach and includes agrarian and environmental movements and has emerged primarily out of the work of the South African Food Sovereignty Campaign and the Cooperative and Policy Alternative Centre. Its members' arguments are rooted in a critique of the capitalist system and the extractive, growth-oriented model that underpins it. In keeping with this, they advance a radical agenda for climate justice in what they call a "*deep* just transition." This centres on systemic alternatives for transformative change, placing climate justice and the energy transitions within a broad vision of development that prioritises principles of social, ecological and intergenerational justice along with social ownership and community-based renewable energy, participatory democracy and well-being. This vision is advocated through key demands set out in their charter, which they presented to parliament to be adopted under Section 234 of the Constitution of South Africa. They have also been advocating for a boycott on financing for the energy transition. To this end they launched a petition to the UNFCCC in 2022 to make ending coal, gas and oil investment a condition for financial support to South Africa. This is on the grounds that the South African government's current energy policies and actions are not consistent with moving away from a fossil fuel-based energy system.

Such a petition challenges the current efforts by the South African state to raise international finance to fund the energy transition. This includes the much celebrated $8.5 billion dollars promised to South Africa as part of Just Energy Transition Partnership signed at COP26 in Glasgow to fund the decarbonisation of the energy sector, and more recently set up the Just Energy Transition Investment Plan (JET IP) which was launched at COP27 in Egypt in 2022 in an attempt to attract international investment finance.

6.4.5. Financing the just transition: Taxation, loans and debt

Movement actors are forwarding a range of arguments regarding financing the just transition and LFFU. These range from ideas around progressive taxation to sanctions and debt cancellation, each underpinned by different assumptions about responsibility and liability. Progressive taxation is widely advocated by the labour movement. COSATU calls for a progressive wealth tax, a resource rent tax, and an environmental damage tax, while SAFTU is explicit is its demands, calling for a corporate tax be implemented at a minimum rate of 48% (SAFTU, 2020a). Such progressive taxation should also be seen as a form of climate and apartheid reparations (Lenferna, 2023). This argument rests on the fundamental link between the apartheid system (and the colonial system more broadly) and its role in enabling the highly polluting and exploitative minerals–energy complex (amongst other industries), which has historically, and continues to rely on the mass exploitation of people and nature. The apartheid system of racialised discrimination, dispossessed black people from their land and thus created a cheap labour force which could serve the capitalist interests of the rapidly expanding white industrial sectors, particularly that of minerals and energy. At the same time the mass resettlement programmes and the creation of the former homelands on suboptimal lands, where people were subjected to severe overcrowding, led to environmental degradation, and more generally underdevelopment. As a result, these spaces and the mostly poor black people that reside there today are at the forefront of the impacts of climate change. This political history has meant that wider debates about climate reparations and compensation from rich countries to poor countries for the loss and damage they suffer at the expense of climate change is contested in the South African context. South Africa being such a massive polluter and still heavily dependent on and actively promoting further fossil fuel extraction, calls into question their legitimacy when it comes to seeking reparations and climate finance from wealthy nations as has been the case in the two most recent COP conferences.

During the COP26 in Glasgow in 2021, the Just Energy Transition Partnership (JETP) was launched, whereby a group of wealthy nations[5] pledged $8.5 billion to help South Africa move to renewables and implement the revised nationally determined contributions (NDCs). The following year during the COP27 in Egypt in 2022, the president launched the Just Energy Transition Investment Plan (JET IP) which aims to raise R1.5 trillion for financing as a further contribution to decarbonise the economy. While these appear to be progressive steps towards realising climate action, the Climate Justice Charter Movement (CJCM) has accused the government of engaging in "double speak" as it talks "green" while it continues to actively engage and promote new fossil fuel extraction projects. Hence, CJCM has called for a boycott of all international financing for South Africa's just transition until it demonstrates its commitment to decarbonising the economy by ending fossil fuel extraction (CJCM, 2022). The secretary of the Climate Justice Coalition highlights that for climate finance to be morally justified and to serve climate justice, it needs to be in line with South Africa's fair share[6] of climate action and must help transform the deeply unjust country and ensure the benefits of such a transition don't just perpetuate existing inequalities and benefit the elite, an inevitable outcome if the current market-driven transition is pursued (Lenferna, 2023). He highlights: "[T]he task of social movements committed to climate justice is to ensure that climate finance is not a tool for neocolonialism, but rather advances true economic, environmental and social justice" (Lenferna, 2023).

On closer scrutiny of the JET IP, the finance secured thus far is primarily concession grants (63%) and commercial loans (18%) with as little as 4% in grants (IEJ, 2022). These funds will be used to "de-risk" investment by the private sector and to facilitate "blending" public finance with private finance to attract more foreign private finance, a model that is considered dangerous by the Institute for Economic Justice as it stands to facilitate profiteering by the financial elite while limiting the possibilities for local economic development (IEJ, 2022). Furthermore, by facilitating privatising electricity generation through independent power producers (IPPs), while the government takes the risks as is currently the case, will only raise the cost

5 This included the United States, France, Germany the United Kingdom and the European Union.

6 This is a calculation developed by the Climate Action Tracker which seeks to help counties quantify their emission reduction targets, in line with the general principles of "highest possible ambition" and "common but differentiated responsibility and respective capabilities, in light of the different national circumstances" laid out in the Paris Agreement (Climate Action Tracker, 2023).

of energy provision and limit access by poor households only exacerbating energy poverty (IEJ, 2022). Local economic development will be compromised because most of the finance raised is earmarked for infrastructure development in the electricity sector, whereas the "justice" element of the JET IP is mostly in the economic diversification and innovation, as well as skills development for which only 0.3% and 0.1%, respectively, was allocated (IEJ, 2022). There is little support for the localisation of renewable energy value chains, while it is in manufacturing, not electricity generation, where jobs stand to be created. The Institute for Economic Justice warns that the JETP and JET IP "risks replacing the ecological crisis with a colossal social crisis" (IEJ, 2022, p. 3).

While the South African government continues its drive to mobilise international finance, driving the country into ever more debt, there is a growing movement mobilising around debt cancellation, both based on the argument of odious debt but also based on the climate debt owed by rich countries to the Global South (see also 5.4). The Debt for Climate movement was co-founded by a South African climate activist in early 2022. While relatively new, the movement has galvinised around the long-standing argument for debt cancellation most notably the Jubilee Debt Campaign to which Debt for Climate is aligned. Their key argument is that the Global North owes the Global South a climate debt, and if this debt is cancelled, this revenue can finance a just transition. "If we call for the cancellation of the financial debt against the climate debt we are returning climate activism into climate actions" (interview with an activist and co-founder of Debt for Climate). He goes on to explain:

> The protest movement is very valuable to popularise and expose and mainstream the arguments, but it doesn't inherently have a mechanism to make the change.... We say if you cancel the debt of R200 billion in interest payments per year, that is almost two times the GDP. So, if we can cancel the debt or at least a portion, then we don't need to take the fossil fuels out of the ground to get the revenue to service the debt. In other words, cancel the debt and we'll keep the fossil fuels in the ground. That is a real proposal that can happen. (Interview with activist and co-founder of Debt for Climate)

The World Bank gave a US$3.75 billion loan to Eskom in 2010, to primarily finance the completion of Medupi, the largest coal-fired power station in the world. The climate impact, local environmental pollution and widely documented corruption surrounding this loan have led growing demands for it to be considered odious debt (Cannard, 2019). The Debt for Climate movement is demanding that odious debt and other debt held by the Global South be cancelled as a means of climate reparations for loss and damage, but also that global financial institutions be democratised so as to ensure odious debt such as that provided for Eskom never be repeated (Bannon & Morgan 2023; Morgan 2022).

6.5. Conclusion

This chapter has demonstrated how since the 2010s there has been an expansion and convergence in social mobilisation for LFFU across South Africa. This is particularly impressive considering the deep and widespread social and economic deprivations currently experienced by the majority. The arguments, approaches and strategies employed by the different actors in this space are diverse. However, despite these differences, many have managed to unite in broad coalitions for climate justice. These actors have found unity in purpose in the Climate Justice Coalition, which shows promising signs of a diverse yet unified counter movement emerging from below. This convergence is largely a result of important strides made by INGOs, NGOs and CBOs who are successfully linking the daily struggles of poor and marginalised communities for basic services and livelihoods to fossil fuel extraction and climate change. Single-issue campaigns and litigation have paved the way for communities to hold government and corporates to account when it comes to procedural grounds for granting mining licenses and assessing potential environmental damage. It has also emboldened communities to call for a greater voice in decision-making around mining more generally. These niche activities have led to the convergence between societal groupings that show positive signs of catalysing a wider regime change. Here particularly the convergence of labour, social and environmental movements under the umbrella of the CJC is a very promising development.

Beyond the important work being done around fighting single-issue cases through the successful partnerships between NGOs and community organisations, significant arguments engage with the politics of the possible and use this as a political moment to shift beyond a just transition that focuses on LFFU, and instead focus on progressive and transformative ideas for a post-capitalist world focused on eco-socialist and feminist alternatives. Here the progressive labour movement and other progressive social groupings such as the CJCM and CJC have become an important force. They have centred capitalism and its extractive and exploitative nature as the driving force behind the everyday

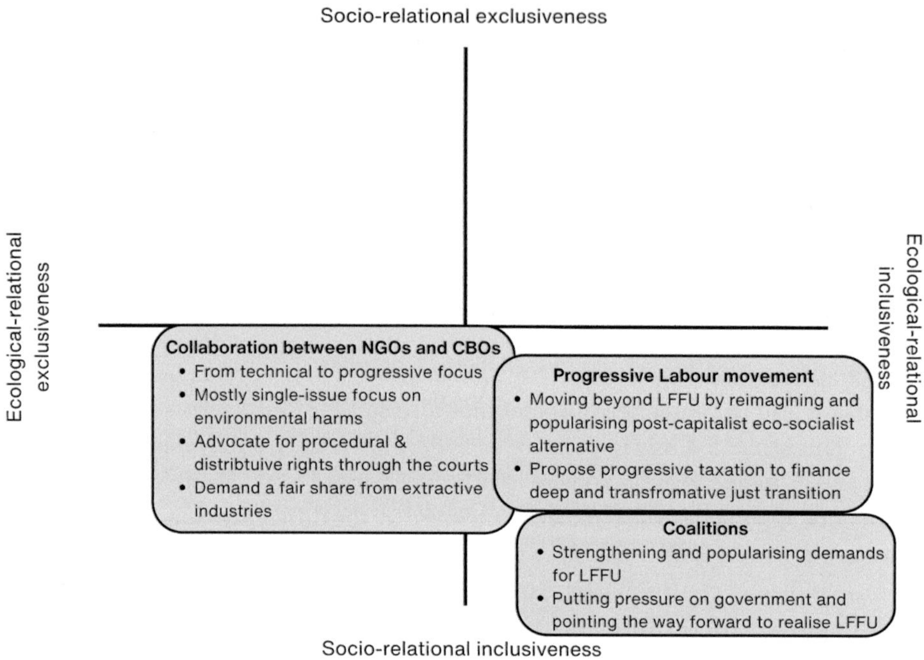

Figure 6.1. South African movements from below and inclusive transitions.

experiences of environmental degradation and social issues. A clear vision of what a post-capitalist society could look like is emerging. The "democratic deep just transition" forwarded by the CJCM and the eco-socialist vision of COSATU, or the ecofeminist vision of WoMin, all centre the capitalist system as the root cause of the social and ecological crisis. They share a common framing of the problem and emphasise a similar vision for a just transition. This extends beyond a technical and social focus of a just energy transition that simply replaces fossil fuel energy generation with green alternatives, while ensuring that labour is reskilled and job losses prevented. These concrete visions and the detailed alternatives that are put forward present an important move towards forwarding the politics of the possible. These alternative visions of society and what development could mean stand to inform what landscape-level transitions could look like (see Figure 6.1).

References

Bannon, M., & Morgan S. (2023, February 26). Cancel odious Global South debt to accelerate climate justice. *Daily Maverick*. https://www.dailymaverick.co.za/

article/2023-02-26-cancel-odious-global-south-debt-to-accelerate-climate-justice/
Bond, P. (2012). *Politics of climate justice: Paralysis above, movement below.* University of Kwa Zulu Natal Press.
Bond, P. (2015). Can climate activists "movement below" transcend negotiators "paralysis above"? *Journal of World-Systems Research, 21*(2), 250–269.
Bond, P. (2018). Ecological-economic narratives for resisting extractive industries in Africa. In P. Cooney & W. S. Freslon (Eds.), *Environmental Impacts of Transnational Corporations in the Global South* (pp. 73–110). Emerald Publishing. https://doi.org/10.1108/S0161-7230201833
Bridle, R., Muzondo, C., Schmidt, M., Laan, T., Viswamohanan, A., & Geddes, A. (2022). South Africa's energy fiscal policies. International Institute for Sustainable Development. https://www. iisd. org/publications/south-africa-energy-subsidies
Budlender, J., Leibbrandt, M., & Woolard, I. (2015). South African poverty lines: A review and two new money-metric thresholds. SALDRU [Southern Africa Labour and Development Research Unity] Working Paper 151. University of Cape Town.
Cannard, J. (2019, September 3). South Africa: Cancel Eskom's odious debt to the World Bank. Committee for the Abolition of Illegitimate Debt. https://www.cadtm.org/South-Africa-Cancel-Eskom-s-odious-debt-to-the-World-Bank
CJC [Climate Justice Coalition]. (2022). Report on the work of the Climate Justice Coalition.
CJC [Climate Justice Coalition]. (N.d.). Uproot the DMRE. https://uprootthedmre.org/
CJCM [Climate Justice Charter Movement]. (2022). Make ending coal, gas and oil investment a condition for financial support to South Africa. Change.org petition
CJCM [Climate Justice Charter Movement]. (2023). https://cjcm.org.za/campaigns
Climate Action Tracker (2023). https://climateactiontracker.org/methodology/comparability-of-effort/
Cock, J. (2019). Resistance to coal inequalities and the possibilities of a just transition in South Africa. *Development Southern Africa, 36*(6), 860–873. https://doi.org/10.1080/0376835X.2019.1660859
COSATU [Congress of South African Trade Unions]. (2011, August). COSATU policy framework on climate change. http://www.cosatu.org.za/show.php?ID=5679.
COSATU [Congress of South African Trade Unions]. (2022). Just transition blueprint for workers 2022: Summary document.
Earthlife Africa Johannesburg and Southern African Faith Communities vs The Ministry of Energy. (2017). Case no. 19529/2015. Western Cape High Court, South Africa.
Earthlife Africa Johannesburg vs The Ministry of Environmental Affairs. (2017). Case no. 65662/16. Gauteng High Court, South Africa.

Faber, T. (2022, November 20). Fracking hell: Outrage as Mantashe calls for proposals for seismic activities in Karoo. *Times Live*. https://www.timeslive.co.za/sunday-times-daily/news/2022-11-20-fracking-hell-outrage-as-mantashe-calls-for-proposals-for-seismic-activity-in-karoo/

IEJ [Institute for Economic Justice]. (2022, November 10). Secretly-negotiated South African "climate finance deal" a gist to private investors while choking local development. IEJ Statement.

Kamanzi, B. (2022, December 6). Komati decommissioning: A spectre due to haunt the just transition. *Amandla, 85/86*. https://amandla.org.za/komati-decommissioning-a-spectre-due-to-haunt-the-just-transition/

Lancaster, L. (2018). Unpacking discontent: Where and why protest happens in South Africa. *South African Crime Quarterly, 64*, 29–43.

Lenferna, A. (2022, January 30). It's not foreign forces—It is the people of South Africa who are fed up with Gwede Mantashe and the DMRE. *Daily Maverick*. https://www.dailymaverick.co.za/opinionista/2022-01-30-its-not-foreign-forces-it-is-the-people-of-south-africa-who-are-fed-up-with-gwede-mantashe-and-the-dmre/

Lenferna, A. (2023, April 24). Does South Africa deserve climate reparations? *Daily Maverick*. https://www.dailymaverick.co.za/opinionista/2023-01-24-does-south-africa-deserve-climate-reparations/

Maggott, T., Mbatha, S., Ceruti, C., Moyo, L., Mporo, A., Ngwane, T., Shezi, C., & Sinwell, L. (2022). Energy racism report: The electricity crisis and the working class in South Africa. Centre for Sociological Research and Practice (CSRP), University of Johannesburg.

Moolman, S. (2022). 2022 update: Eskom tariff increases s inflation since 1988 (with projections to 2024). PowerOptimal website. https://poweroptimal.com/2021-update-eskom-tariff-increases-vs-inflation-since-1988/

Morgan, S. (2022, October 13). The quickest way to reduce Eskom's balance sheet—Declare odious debt on Madupi and Kusile. *Daily Maverick*. https://www.dailymaverick.co.za/opinionista/2022-10-13-the-quickest-way-to-reduce-eskoms-balance-sheet-declare-odious-debt-on-medupi-and-kusile/

Mwenda, M., & Bond, P. (2020). African climate justice: Articulations and activism. In B. Tokar & T. Gilbertson (Eds.), *Climate justice and community renewal: Resistance and grassroots solutions* (pp. 108–127). Routledge.

The Presidency. (2022). *South Africa's Just Energy Transition Investment Plan (JET IP) for the initial period 2023–2027*. Republic of South Africa.

SAFTU [South African Federation of Trade Unions]. (2020a). Notice to Nedlac about possible protest actions. Section 77(1)(b).

Sangari, K. (2002). *Politics of the possible: Essays on gender, history, narratives, colonial English*. Anthem Press.

Satgar, V. (2015). A trade union approach to climate justice: The campaign strategy of the National Union of Metalworkers of South Africa. *Global Labour Journal, 6*(3).

South African Government. (2011). *National climate change response: White paper.* Department of Environmental Affairs and Tourism.

UNFCCC [United Nations Framework Convention on Climate Change]. (2015). The Paris Agreement. United Nations. https://unfccc.int/sites/default/files/english_paris_agreement.pdf

7. Finance and LFFU

Arthur Rempel, Joyeeta Gupta, Nina Brander, Clara McDonnell, Anne Elizabeth Politsch, Graham Tennant Green, Tjeerd Harkema, Emma McGarthy and Adrien Tofighi-Niaki

Abstract

Financial actors play a key role in the fossil fuel sector and thus can be expected to play a role to reduce greenhouse gas emissions. This chapter focuses on global actors such as investors and producers and their relations with the Global South. It examines philanthropic foundations, pension funds, investor initiatives and export credit agencies. Our research shows that although these actors have a critical role, they are not taking any meaningful action through LFFU. Although divestment is underway, it is doubtful if that leads to LFFU as new shareholders emerge. In the meanwhile, for a long time, the export credit agencies have been financing fossil fuel projects in developing countries and have only recently committed to phasing that out.

Keywords: fossil fuel sector, divestment, financial actors, Global South, Global North

7.1. Introduction

Given that fossil fuel reserves, resources and assets may cumulatively aggregate to about $200 trillion (Linquiti & Cogswell, 2016), large (and global) investors and financiers play a strategic role in LFFU. Despite this, there are critical gaps in knowledge regarding the activities and contributions of large investors. Hence, this chapter addresses the question: What role do global philanthropic foundations, pension funds, investor coalitions and export credit agencies play in both financing the fossil fuel sector and contributing towards climate change mitigation? With the resurgence of

---

Gupta, J., Hogenboom, B., Rempel, A. & Olofsson, M. (eds), *Leaving Fossil Fuels Underground: Actors, Arguments and Approaches in the Global South and Global North.*
Taylor & Francis Group, 2024
doi: 10.5117/9789048560370_CH07

neoliberal economic policies, including increased deregulation, the retreat of government public provision and the prioritisation of shareholder value, investors are now positioned to have a significant impact on climate change mitigation. They control increasing shares of global wealth and provide critical finance and have the capacity as shareholders to influence company decision-making, though this influential role is understudied with respect to fossil fuels. Hence, we examine investors' statements, commitments and activities to interrogate how they may directly or indirectly contribute towards LFFU. Relational inclusiveness is often challenged by the ways in which producers and investors engage in governance processes, allocate finance, promote deregulation at national and international levels, engage in self-governance, and arrange public private partnerships. Through this chapter, we examine how the main purposes driving investors, from profit-incentives to investing for "doing good," can complicate, counter, and undermine climate mitigation.

We focus on global actors (investors and producers, see 2.3.3) and their relations with the Global South, building on our theoretical approach. We discuss philanthropic foundations (see 7.3), pension funds (see 7.4), investor initiatives (see 7.5) and export credit agencies (see 7.6) before drawing conclusions (see 7.7).

7.2. Building on the analytical framework

This chapter assesses the key resources that are being channelled into the fossil fuel sector globally. In order to do so, we have focused on four key actors—philanthropic organisations, pension funds, investor initiatives and export credit agencies. We have used different methods to understand each of these actors. Philanthropic foundations' contribution to climate change mitigation through LFFU was explored using a mixed methods research design. This included a literature review, eight semi-structured interviews and content analysis of 26 annual reports (2018–2020) of philanthropic foundations that spanned the US, Europe and the UK. Pension funds and their role in financing the fossil fuel industry were explored by analysing three US state pension funds and three UK local government pension schemes using document analysis, interviews and an online survey. Investor initiatives related to climate change mitigation were inventoried through analysis of the annual reports, sustainability reports and websites of a sample of 70 of the world's largest institutional investors. A qualitative content analysis of initiatives' websites and annual reports was used to examine their potential contribution to LFFU. Critical frame analysis, supported by a content analysis of 55 media reports, was then used to examine one initiative, Climate Action 100+, as a case study.[1] Export credit agencies' role in financing the fossil fuel industry was analysed by focusing on Dutch, Canadian and UK agencies and their relation to climate change mitigation, with a specific focus on the impacts of the Dutch export credit policy on Nigeria. The analysis included content analysis and semi-structured interviews.

7.3. Philanthropic foundations

Philanthropic foundations (from here on "Foundations") are difficult to define (Anheier & Daly, 2006; Daly, 2012; Ulbert & Hamm, 2011) and have varied characteristics (Daly, 2012; Anheier & Daly, 2006; Morena, 2016). What binds them is that they are mostly independent, private (or community) organisations with their own financial assets, and which are self-governing, non-state actors that ostensibly serve a public purpose (Anheier & Daly, 2006; Morena, 2016). Private and independent Foundations generally receive their funds from a company, individual or family (Anheier & Daly, 2006; Morena, 2016); they are mostly governed by a board of trustees (Koushyar et al., 2015). Community Foundations generally get money from their community but otherwise function as others. Their role in the fossil fuel sector is scarcely covered in the literature.

In addition to being self-governing and non-profit entities, contemporary Foundations are also usually asset based (Anheier & Daly, 2006; Morena, 2016). In fact, total assets under management by US-based Foundations have increased by almost 100% recently, from $480 billion in 2003 to $870 billion in 2014 (Foundation Center, 2014; Hammack, 2006); this puts Foundation assets under management on the same order of magnitude as fossil fuel assets (liquid) managed by OECD pension funds in 2019 (Rempel & Gupta, 2020; see 7.3). This growth is typically attributed to the increasing prominence of public–private partnerships in provisioning public goods (Adloff, 2015; Ulbert & Hamm, 2011). Like pension funds and other institutional shareholders, Foundations invest their mammoth funds in diversified portfolios in hopes of earning a profitable return on their investment, and subsequently allocate their profits to their charitable, "do good" activities; during the same 2003–2014 period that saw US Foundation assets increase twofold, charitable contributions by Foundations for international

1 Results of this research have been discussed in more depth in McDonnell et al. (2022).

development-related purposes also experienced a substantial spike (Ulbert & Hamm, 2011; Vogel, 2010).

Increasingly, Foundations have been funnelling their charitable activities to allegedly take climate action through "greenhouse gas mitigation." Through various mechanisms (including grants, programmes, prizes and impact investments) Foundations sought to promote renewable energy and energy efficiency projects, which are two ubiquitously discussed LFFU approaches (e.g. Baldwin et al., 2020; Green & Denniss, 2018; Johnstone et al., 2017; Healy & Barry, 2017; Lazarus et al., 2015; Erickson et al., 2018). Although, it should be noted that the effectiveness of both approaches vis-à-vis LFFU has been put into question by the most recent scholarship (Rempel & Gupta, 2022). Only five of the sampled Foundations explicitly acknowledged the climate emergency,[2] though *none explicitly aimed to align their climate action with a policy of LFFU.* Furthermore, the sampled Foundations jointly managed assets worth $182 billion in 2018, of which only 6% ($11 billion) was allocated for charitable programmes, whereas the overwhelming majority ($171 billion, 94%) remained as investments, suggesting that Foundations act more like glorified savings accounts or asset managers than the societal benefiting organisations that they market themselves as.

Moreover, questions regarding minimal regulation, the source of their capital and investment decisions (in addition to their lacklustre climate action) render Foundations subject to additional scrutiny and criticism. The primary regulation of Foundations focuses on the maintenance of their tax privileged status and their adherence to financial laws—whereby they cannot engage in outright financial malfeasance, such as embezzlement, fraud etc., and they must file yearly tax forms (Thelin, 2014a, 2014b). With such limited regulation, accountability and transparency are essential (Anheier & Daly, 2006; Hesselmann, 2011; Morena, 2016). However, there is historically an alarming lack of accountability and transparency within the philanthropic sector (Anheier & Daly, 2006; Hesselmann, 2011; Morena, 2016). Some argue that the charitable activities engaged in by Foundations are undertaken only to avert public criticism, diverting attention from what is essentially a legally legitimised system of tax avoidance and sheltering (Anheier & Leat, 2013). Moreover, and perhaps most problematic, much Foundation start-up capital originates from activities which themselves create and exacerbate the issues Foundations dedicate their activities towards addressing (Berger & Przyrembel, 2019; INCITE! Women of Color

2 Charles Stewart Mott Foundation, Esmée Fairbairn Foundation, IKEA Foundation, the Kresge Foundation, and Wellcome Trust.

against Violence, 2017). By channelling their fortunes into Foundations, it is alleged that elite actors work to obscure the harmful origins of their capital, conveniently avoiding tax and maintaining control over their assets (Fisher, 1983; Hesketh, 2017; Kourula & Delalieux, 2016).

The origins of Foundations' capital are controversially and paradoxically linked to the fossil fuel sector. Eleven Foundations from our analysis revealed direct links between their portfolio investments and energy-intensive assets, of which five explicitly denoted links to fossil fuel-related assets. The Gulbenkian Foundation (based in Lisbon), for example, were the original major shareholders of Partex Oil and Gas Corporation, an oil and gas exploitation company with Turkish and Panamanian roots dating back to the 1980s (Partex Oil and Gas, n.d.). In addition to the Gulbenkian Foundation, two others (the Richard King Mellon Foundation and the Shell Foundation) also came to existence through oil and gas exploitation over the last decades. Although Foundations often self-regulate by declaring "areas of non-investment"—i.e. sectors and industries in which they pledge to refrain from investing—we find that these areas are strictly limited to historically controversial industries, like tobacco and arms trade, and omit fossil fuel extraction and production. Three respondents from our semi-structured interviews explained that they "had no idea if their Foundation helps investment in fossil fuels," while two others explained that although their Foundations did not manage fossil-related assets, there are no policies or guidelines in place to prohibit such investments.

In 2019, however, in a bid to align themselves with more "sustainable" practices, the Gulbenkian Foundation divested its Partex shares to PTT Exploration, a Thailand-based oil and gas conglomerate, for over $600 million (TPN/Lusa, 2019)[3]; moreover, our interview respondents revealed that divestment was actively pursued by at least one Foundation, another considered "divestment an eventual outcome" of aligning with more sustainable investments, and three others were unable to confirm whether their Foundations practiced fossil fuel divestment. This divestment technique is not only ineffective but also inequitable and unjust as it absolves the leaders of the Gulbenkian Foundation from any and all accountability for the 40 years of oil and gas exploration that they personally and directly executed, reallocating the responsibility and burden of LFFU to other investors (see 7.4). Only one Foundation (Esmée Fairbairn) actively engaged with companies in which they invest on climate issues, pushing them to "set

3 Note that the other Foundations did not openly disclose the details vis-à-vis their fossil-related investments despite confirming them.

credible science-based targets in line with the long-term climate goals set out in the Paris Agreement." Five interview respondents confirmed a general appeal with engagement, though research suggests that engagement too has historically been ineffective for LFFU (Rempel & Gupta, 2020).

This analysis has shed light on a dichotomous set of misalignments that severely puts into question the extent to which Foundations are aligning with global climate objectives. First and foremost, the financial capital with which Foundations operate their businesses has prominent (past and present) linkages to the fossil fuel sector. Their role as shareholders is problematised through ineffective and relationally exclusive practices like divestment, altogether indicating that from a financial standpoint, Foundations can be characterised as relationally exclusive institutional shareholders that are happy accruing profits for their "charitable programmes" through climate-deteriorating assets. On the other side, the very nature of their charitable programmes raises questions about the effective allocation of charitable funds for addressing climate change; Foundations often opt for financing "green" projects rather than e.g. allocating funds to decommission existing fossil fuel infrastructure (Sovacool & Geels, 2016; see 9.2.2); and only allocate a miniscule fraction of their assets for said programmes, with the vast majority remaining as investment. Both sides of the coin—i.e. their financing and charitable actions—suggest that Foundations are completely misaligned with the Paris Agreement's objectives, and their role in mitigating against the climate emergency must be completely reimagined.

7.4. Pension funds

Like Foundations, pension funds (from here on "Pensions") are also institutional shareholders and invest in a diversified portfolio of assets; however, unlike Foundations, Pensions hold the primary purpose of accruing profits so as to pay retirement instalments (Sarang, 2015; Amadeo, 2021) rather than engaging in "charitable activities" (which, again, we question; see 7.3). Because of their mammoth sizes, Pensions are able to accrue interest quickly; between 1981 and 2015, global assets under Pension management increased by 3,262% (OECD, n.d.), and by 2019 they aggregated to some €41 trillion (OECD, 2020b). In essence, Pensions are massive savings accounts set up by corporations, governments and employee groups (Hinz et al., 2010) that are jointly responsible for managing circa 10% of net global wealth as of 2021 (Williams, 2021).

Furthermore, unlike Foundations and other for-profit investors (private equity, hedge funds), Pensions—and the asset managers that manage their portfolios on their behalf—have a fiduciary duty to comply with, denoting a legal obligation to act in the best interest of their beneficiaries (Schanzenbach & Stikoff, 2020). Historically, this fiduciary duty has been understood as an obligation to invest in the best interest of their beneficiaries, with a primary focus on earning returns while minimising risk (Rempel & Gupta, 2020). However, with the climate emergency looming, the scope and priorities of fiduciary duty may be subject to change; for many, it's not just a matter of having more funds with which to retire, but also retiring in a world capable of sustaining life.

Pension portfolios have been—and continue to be—highly invested in fossil fuel assets. For instance, in 2014, the Local Government Pension Scheme in the UK had invested over £12 billion in liquid assets (i.e. common shares and convertible bonds) pertaining to fossil fuel producing firms (Fossil Free UK, n.d.). In 2019, a sample of 15 Pensions (with total liquid assets summing to €2 trillion) held almost €80 billion in liquid assets pertaining to coal, oil and gas exploration and production companies, or roughly 4% of total assets, suggesting that Pensions managed between €238 and €828 billion in fossil assets globally (Rempel & Gupta, 2020). Despite these assets shrinking by some 40% during the COVID-19 "pancession" (pandemic + recession) (Rempel & Gupta, 2021), Pension fossil fuel assets continue to dominate in 2021: Norway's Government Pension Fund *alone* managed $49 billion (as of July 2021) in liquid assets pertaining to a sample of 20 fossil producing firms; major Dutch Pensions ABP and Zorg & Welzijn jointly managed roughly $3 billion in the same small sample; and over 10% of Sasol's (major South Africa-based oil and gas multinational) shares ($514 million of $4.9 billion) were managed by South Africa's Government Employee Pension Fund in July 2020.[4] As major shareholders, Pensions play a key role in governing prominent fossil fuel conglomerates by e.g. voting for shareholder resolutions at annual meetings, instituting new board members and actively governing these firms more broadly (Anabtawi & Stout, 2008).

Pensions face a series of dilemmas with respect to these fossil-intensive investments. The first is underpinned by the *stranded asset risk* (Bos & Gupta, 2018) that fossil-intensive assets pose to Pensions (and all shareholders more broadly). In the event that climate policies are tightened so as to accelerate the coal, oil and gas phase-out to comply with at least a 1.5 °C average temperature rise (IPCC, 2021), these fossil-intensive assets may be

4 Original analysis; see methods in Rempel (2023, p. 79).

severely devalued and left "stranded" on the balance sheets of Pensions as the "carbon bubble" bursts (Carbon Tracker Initiative, 2013) and the global economy experiences a dip in discounted global wealth worth trillions (Mercure et al. 2018). By holding onto fossil assets, Pensions expose their beneficiaries to a (potentially severe) risk through their fossil-intensive assets—but what should they do about it?

One option is to *divest* from these assets—generally defined as selling assets with a particular social, financial or political objective (Finley-Brook & Holloman, 2016). By divesting, Pensions would rid themselves of the aforementioned stranded asset risks as they pass on the risk-prone asset to another (potentially more risk-tolerant) investor. However, by doing so, Pensions would simultaneously be absolving themselves of responsibility and accountability to take effective climate action as the burden of doing so is reallocated (along with the stranded asset risk) to the new, risk-tolerant buyer (Gupta et al., 2020). Activists typically adopt divestment as a core strategy in their climate agendas—manifested through the fossil fuel divestment movement, which has compiled some $11 trillion in *pledged* divestments from 2010 to 2019 (Fossil Free, 2019). High-profile investors, including the Dutch Pensions PME and ABP, announced in 2021 that they would divest their fossil-intensive assets. However, even on such a global level, divestment is unlikely to yield any real financial consequences for fossil fuel producing firms (Bergman, 2018; Ansar et al., 2013) other than *marginally* increasing their cost of capital (Gunningham, 2020) seeing as an "international law" mandating sector-wide divestment doesn't (and probably will never) exist (Matikainen & Soubeyran, 2022). As such, divestment is better justified on either moral/ ethical grounds ("it is bad to profit from companies that propel the climate emergency") (Grady-Benson & Sarathy, 2016) or its stigmatisation of the fossil fuel industry (Sarang, 2015; Schneider, 2014; Ansar et al., 2013).

Rather than divest, Pensions could leverage their shareholder power and engage the fossil companies in hopes of altering the firm's business practices, though this raises a second dilemma. Fossil fuel producers specialise in producing hydrocarbons (fossil fuel), not electrons (energy), even though some major oil and gas multinationals are trying to rebrand themselves as more "sustainable" firms in light of the climate emergency (e.g. French oil and gas giant Total changed its name to TotalEnergies in June 2021 to place emphasis on the "green power" it produces in addition to its unabated fossil fuel business [Takahashi, 2021]). A proactive and successful engagement with these firms would require a quick and drastic fettering of their hydrocarbon production, thereby devaluing the firm's assets and solidifying the aforementioned stranded asset risk on the Pension balance sheets—violating traditional interpretations of fiduciary duty. For engagement to succeed, our understanding of fiduciary obligations must be completely modernised and reimagined in the context of the climate emergency (Rempel & Gupta, 2020); effective climate action via LFFU implies a gargantuan cost, so who will incur it?

Moreover, evidence thus far suggests that Pensions are not yet ready to grapple with the engagement dilemma; in a sample of 15 prominent Pensions, only one instance of proactive LFFU engagement was identified in their 2019 sustainability reporting; few other fossil-related engagements were detected, but these were reactive in nature and limited to discussing oil spills or corruption allegations (Rempel & Gupta, 2020). Without a proactive and progressive push to LFFU, engagement efforts will continue to, at best, distract from effective climate action and, at worst, fuel the climate emergency by allowing fossil fuel production to run rampant at the global stage.

Dilemma aside, past engagement efforts have been incredibly time-consuming. In 2000, a mix of stakeholders (Greenpeace, SRI funds, public interest associations and several individual investors) submitted a shareholder resolution with BP asking the fossil fuel company to "halt the development of the Northslope field in Alaska and redistribute the investment to the BP Solarex (solar energy) division" (O'Rourke, 2003, p. 234). The proposal was voted on by BP's shareholders on 13 April 2000, at the annual general meeting (AGM), where 13.5% of shareholders, representing roughly 1.5 trillion shares, voted "yes" in support of the proposal (O'Rourke, 2003). After the 2000 AGM, Greenpeace stated that BP had in fact acknowledged the global climate change crisis and the fossil fuel company's role; however, there remained concern over the fact that BP has not disclosed to shareholders how the fossil fuel company would make the transition to renewables, as, at the time, 99.9% of BP's investments were in gas and oil (O'Rourke, 2003). Thus, in 2001, another shareholder resolution was submitted, however, it was excluded by BP "based on a legal technicality" (O'Rourke, 2003, p. 235). In 2002, the World Wildlife Fund (WWF) led another campaign targeting BP's climate change policy, and subsequently submitted another shareholder resolution in collaboration with other NGOs, SRI and a large institutional investor collective, asking for climate risk analysis and climate disclosure (O'Rourke, 2003). This multi-year instance speaks to the lethargy of the engagement process, and as such, perhaps the engagement mechanism is not well-suited to address the urgency of combatting the climate emergency.

All in all, an analogously dichotomous conclusion to that of Foundations is reached (see 7.3); given their role as fossil fuel investors vis-à-vis the liquid assets managed in fossil fuel producing firms, Pensions are in a prominent position to take action to LFFU, but this may come at a *financial* cost to their beneficiaries (i.e. pensioners) (despite simultaneously yielding *non-financial* benefits to said beneficiaries in the form, e.g. mitigated climate impacts). Managing these financial losses is imperative to adequate and inclusive climate action but would require completely revamping the paradigm in which Pensions (and institutional investors in general) operate. To the best of our knowledge, no Pensions have experienced such a shift in paradigm and begun employing such actions, but it could take as little as one niche instance to inspire a global movement and eventually (inclusively) reshape the Pension landscape.

7.5. Investor initiatives

The effectiveness of shareholder engagement (see 7.4) is frequently dependent on mobilising a collective of large shareholders. While these collectives may come together on an ad hoc basis, Pensions and other large asset owners and asset managers are increasingly active in formalised coalitions or initiatives. Institutional investors have been forming coalitions to strengthen their capacity to influence corporations on social and environmental issues since the 1970s, with the intent to "purposively steer (i.e. govern) the behaviour of market actors (i.e. corporations and investors) through the broad range of tools at their disposal" (MacLeod, 2009, p. 34). While collective action can frequently be hampered by "free riders" who benefit from the collective efforts without putting in the individual work (Kruitwagen et al., 2017), investor initiatives (from here, "Initiatives") have been considered a way to bypass these challenges. By functioning as "enabling organisations" they can enable cost sharing among investors, reduce the costs of coordination, and maintain continued pressure on targeted corporations (Gond & Piani, 2013; Kruitwagen et al., 2017). Since the 2015 Paris Agreement, increasing numbers of high-profile Initiatives have emerged, created specifically to address climate concerns. The large-scale investor backing of emerging Initiatives has led some (e.g. Henderson, 2020) to consider them the solution to governing fossil fuel companies and bringing them in line with climate targets.

While the primary aim of many of the Initiatives is to facilitate collective investor engagement on climate issues and with fossil fuel companies (most notably, Climate Action 100+), Initiatives have also taken up a broader range of climate activities. These include providing tools and frameworks for investors to evaluate the exposure to carbon risk and climate risk of their own portfolios or companies they invest in (e.g. Task Force for Climate-Related Disclosures; Transition Pathway Initiative), and to assess emissions reduction targets (e.g. Science Based Targets Initiative). Others, such as the Net-Zero Asset Owner Alliance (NZAOA), commit investor members to emissions reduction targets for their own portfolios in addition to facilitating engagement. Though they are predominantly headed by institutional investors, many of these groups receive the support and backing of multilateral institutions (e.g. the UN, G20) or various NGOs (e.g. WWF). Initiatives have been instrumental in shifting attention to climate issues within the financial sector. Initiatives have been able to unite a critical mass of investors behind their calls for climate action; the most relevant Initiatives working on climate issues are backed by investors who control assets worth between $500 billion and $106 trillion.[5] This large-scale attention to climate from investors has significantly contributed to "mainstreaming" climate conversations within the finance sector and have normalised requests for alignment with 1.5 °C scenarios and adoption of net zero emissions by 2050 targets. They have been able, through Initiatives like Climate Action 100+, to ensure that systematic and continuous investor engagement on climate is taking place at the top polluting companies globally.

However, while the emergence of large-scale, high-profile Initiatives has provided a platform with huge potential for effective shareholder engagement with fossil fuel companies, the policies, organisational structure, and actions of Initiatives undermine this potential. While most call for alignment with the Paris Agreement and the need for net zero targets, very few have released a clear and explicit position on fossil fuels. Their methodologies for tracking and assessing compliance with net zero goals still leave various gaps open for ongoing fossil fuel production. For example, both Climate Action 100+ and NZAOA rely on calculations that measure emissions intensity, rather than absolute emissions (CA100+, 2020; NZAOA, 2021), the former typically defined as carbon equivalent emissions per unit of energy produced or consumed (CO_2e/MJ). With this metric, larger companies with diversified businesses can lower their carbon intensity by, for instance, expanding production in various "low-intensity" sectors (e.g. solar PV production) while *still increasing their fossil fuel production and/or consumption.* Oil and gas companies are notoriously adopting this deceptive technique in their carbon accounting as they attempt to rebrand themselves (see 7.4) as

5 Original analysis; see 7.2 as well as McDonnell et al. (2022) for more extensive discussion of methodology.

becoming more environmentally conscious, such as TotalEnergies' (2020, p. 4) plans for "renewable power surging to 15% of [their] sales versus 35% for oil products and 50% for natural gas" to "achieve a 15% reduction in [their] carbon intensity" while oil and gas continue to make up 85% of their total sales. Moreover, ambiguous positions on carbon offsets or negative emissions technologies also present the danger of permitting extended deadlines for fossil fuel production and use. Initiatives demonstrate the tendency for "ambiguity aversion," characteristic of investors in general, a resistance to make decisions without sufficient data to back them up (Ameli et al., 2020). Within climate Initiatives, this translates to a hesitancy to adopt ambitious policies or positions, instead limiting actions to what can be measured or achieved with currently available methodologies.

The institutional structure of Initiatives along with the expectations of their members also present barriers to their effectiveness in influencing LFFU. There are many benefits to investors to join Initiatives (knowledge sharing, access to tools and research, boosting their credibility on climate issues), but very few barriers to entry. Most Initiatives have minimal requirements for investor members to comply with. For example, Climate Action 100+ requires investors to join the engagement team for one of the 167 target companies but has no criteria for the type of action investors take. This has resulted in investors joining onto the initiative, but then acting in ways that clearly contradict the aims of the Initiative. For example, the asset manager BlackRock, after joining Climate Action 100+ in early 2020, proceeded to vote for only 11% of climate-related shareholder resolutions that year. Climate Action 100+ also has no publicly available criteria to assess the effectiveness, progress or success of their engagements. The pace and content of engagements is left up to the discretion of each individual engagement team. Without clearly delineated goals, timelines, and sanctions for noncompliance, the Initiative risks interminable, slow-paced engagements with only incremental wins. This is reflected in Climate Action 100+'s track record thus far, which has achieved commitments to disclosure or net zero "ambitions" from fossil fuel companies, without any concrete commitments to reduce production (CA100+, 2020). As Van Baal and Ashurst (2021) point out, with a backing of $68 trillion in assets under management, Climate Action 100+ is now bigger than Big Oil. However, if the Initiative doesn't have the capability to harness the backing of those investors in any meaningful way, it (problematically) distracts from meaningful Paris Agreement alignment, while creating the façade that effective climate action is being undertaken by the world's richest and most powerful asset managers.

7.6. Export credit agencies

Export credit agencies (ECAs) differ slightly from Foundations, Pensions and Initiatives vis-à-vis financing fossil fuels. ECAs are often a financial arm of a national government (therefore they are public finance institutions [PFIs]); every industrialised country has at least one ECA (Gupta et al., 2020), while "some have multiple institutions that provide different kinds of export finance, as with China, Japan, and Korea" (Tucker & DeAngelis, 2020, p. 18). ECAs serve the sole purpose of exporting a domestic business/industry overseas "by providing either capital flows to exporters" (i.e. loaning in the same way a bank would), "or insurance to cover uncertainty" and de-risk projects for the domestic exporter (Gupta et al., 2020, p. 309). For instance, the mission of UK Export Finance (UKEF) is "to ensure that no viable UK export fails for lack of finance or insurance while operating at no net cost to the taxpayer" (UKEF, n.d.); it "fill[s] gaps in the private sector's provision of finance and insurance," becoming "involved in transactions where there are risks which the commercial market will not accept without ECA support" (House of Commons, 2019). ECAs will typically "step in to facilitate bigger project deals once the commercial and development banks have acquired part of the funding and need additional capital or insurance."[6]

Under the Export and Investment Guarantee Act of 1991, UKEF must allocate £50 billion annually to support UK-borne exports and cannot discriminate across sectors or between classes. As such, the fossil fuel sector has been a hotspot for ECA finance. Between 1996 and 2001, G8 nation ECAs pumped $419 billion into oil and gas development (Maurer, 2002), and more recently, from 2016 to 2018, ECAs from G20 nations funnelled roughly $32 billion annually to global fossil fuel projects (DeAngelis & Tucker, 2020). Between 2013 and 2018, UKEF allocated £2.5 billion for fossil fuel projects compared £104 million for renewable projects, jointly equivalent to over 20% of its total spending; through these flows, UKEF played a critical role in developing the Cape Three Points integrated offshore oil and gas project in Ghana [Petroleum Commission Ghana, n.d]). Although UKEF has pledged to cease all support for coal exports, "nothing similar for oil and gas" has arisen because "the UK is still an oil and gas producer in its own right."[7] The UK's official policy is "to maximise economic recovery" of oil and gas resources, indicating a lack of willingness by the UK government to reduce oil and gas support (Hinson et al., 2020). Moreover, EXIM (the Export-Import Bank of

6 Expert Interview.

7 Expert Interview.

the US) authorised a $5 billion loan in 2019 to finance a liquefied natural gas pipeline in Mozambique (EXIM, 2019), the largest in the bank's history. Dutch ECA Atradius allocated €1.76 billion for fossil fuel development in 2018 (Both Ends, 2019) completely overshadowing the Dutch Fund for Climate Development's €160 million for climate change mitigation from 2019 to 2023 (DFCD, n.d.). Finally, Canada's ECA allocated CAD 7 billion *annually* to overseas fossil fuel projects from 2012 to 2017, compared to a humble $530 million per year for climate change mitigation from 2015 to 2020 (Government of Canada, 2015).

ECAs typically provide capital or insurance across the entire production supply chain, depending on the expertise of the exporting business, meaning that support for fossil fuel projects may be roundabout and therefore somewhat masked:

> Different countries have companies that play a niche role in the coal and oil & gas sectors; Swedish ECAs, for example, are not investing in coal-fired power plants in South Africa, but heavily in the supporting infrastructure like coal transport through the advanced Swedish trucking industry.[8]

Moreover, German (Euler Hermes) and French (COFACE) ECAs played a monumental role in financing South Africa's Medupi and Kusile 4,800 MW coal-fired power stations (see 7.5) by insuring a combined $3.2 billion syndicated loan (with nine French commercial banks and three German counterparts) to support German firm Hitachi Power Europe—who was contracted to engineer the power plant boiling units—and French firm Alstom—contracted to engineer the plant's steam turbines (Bank Track, n.d).

UKEF indirectly supported fossil generation by guaranteeing an $850 million gas-intensive infrastructure project spearheaded by General Electric, while Dutch ECA Atradius allocates ample funds to support home-grown firms like Damen, a supply vessel and ship producer. Damen's vessels and ships operate in the value chain of oil and gas production, providing offshore support services for oil and gas exploration and extraction processes in Nigeria. Moreover, Damen's vessels and ships support the construction and decommissioning of offshore oil and gas drilling rigs (Damen, 2018; Damen, n.d.). Insiders in the Nigerian oil and gas sector stressed the significance of vessels for the operationality of offshore rigs. For example, the importance of supply vessels derives from logistical services (i.e. transferral of technical staff and containers with goods and construction materials to offshore

8 Expert Interview.

sites) that are needed to enable oil and gas exploration and extraction, and the construction of drilling rigs. Patrol vessels also enable exploration and extraction through the provision of security services against external threats, such as piracy. Thus, Damen's vessels and ships are significant to the ability of integrated oil companies to extract oil and gas in Nigeria. Altogether, ECAs catalyse global coal, oil and gas production—particularly in the Global South—through direct *and* indirect financial support in the form of direct capital *or* insurance & guarantees, rendering them absolutely essential players for LFFU.

The OECD's "Arrangement on Officially Supported Export Credits" (OECD, 2020a) (from here on "Arrangement") is the only existing multilateral framework for ECA regulation; it is a "gentlemen's agreement" (i.e. relying on the integrity of the OECD-member ECAs in order to be upheld) with "the main purpose ... to provide a framework for the orderly use of officially supported export credits" and "foster a level playing field ... in order to encourage competition among exporters based on quality and prices of goods and services rather than on the most favourable officially supported terms and conditions" (OECD, 2020a, p. 10). The Arrangement spells out regulations for six economic sectors, two of which are relevant for LFFU: "renewable energy, climate change mitigation and adaptation and water projects" and "[c]oal-fired electricity generation projects." These regulations include criteria that ought to be met for forthcoming projects, like "low to zero carbon emissions, or CO_2 equivalent, *and/or* in high energy efficiency" (OECD, 2020a, p. 94, author's emphasis). In terms of coal-fired power, the Arrangement places no restrictions on ECA finance for "ultra-supercritical plans" (with steam pressure greater than 240 bar and steam temperatures of 593 °C—or emitting less than 750 g CO_2/kWh) and places partial restrictions on "supercritical" and "subcritical" coal-fired technologies, which are less efficient than their ultra-supercritical counterparts. One interviewee characterised the "practical difference between ultra-supercritical and supercritical" units to be "almost entirely trivial in the global climate change context" and identified these conditions as "massive loopholes that ... give investors the green light to continue financing coal as if nothing were different." Another echoed this, claiming that the "OECD guidelines have huge loopholes ... [so that] you can almost support any coal power plant depending on how creative you are," and given the lack of explicit restrictions for oil and gas finance, it is not particularly surprising that ECAs have continued to perpetuate fossil fuel production across the globe.

A final regulatory framework meriting scrutiny is the OECD's Common Approaches for Officially Supported Export Credits and Environmental and Social Due Diligence (from here on "ESDD"), which was adopted in 2012 and revised in 2016. This non-legally binding document urges members to "consider the positive and negative environmental and social impacts of projects ... and the environmental and social risks associated with existing operations in their decision to offer official support for export credits" (OECD, 2016, p. 3). However, the ESDD "only covers projects over a financial threshold"; UKEF has been supporting projects well below that threshold by constructing "repayment terms for less than two years.... [T]hat's how they've been slipping through a lot of the oil and gas stuff, they have been doing it on under two-year agreements."[9]

In contrast to the more indirect role that institutional shareholders (discussed in 6.3–6.5) generally play in supporting the fossil fuel sector, ECAs have had an imperative role in directly driving unabated fossil fuel extraction and production across the entire supply chain across the globe, particularly in the Global South as governments from the North seek to export their climate-deteriorating businesses. But like the previous actors, ECAs are also faced with a dilemma. Eradicating ECA support for coal, oil and gas projects (whether through loans or guarantees) may hamper prospects of securing commercial finance for fossil fuel production as projects are deemed too risky, aligning with global climate objectives; however, this de facto implies slashing trade and exports from a particular country, which ministers of trade will certainly protest. Overcoming this dichotomy—and the lacklustre construction of the OECD frameworks—may hold the key to reinventing the role that ECAs play in LFFU.

7.7. Conclusion

In spite of their critical role as either major shareholders or direct financiers of fossil fuel production, Foundations, Pensions, Initiatives and ECAs are slow to show signs of undertaking any meaningful climate action through LFFU, although some action is visible since this research was undertaken. The shareholder dimension is plagued with a stagnated debate on whether to divest or engage, neither of which show empirical signs of timely effectiveness and may do nothing more than distract from climate action and divert the responsibility and accountability of LFFU elsewhere (in the case of divestment). On the financier end of the spectrum, unabated and manipulative support by ECAs for exporting fossil businesses, through

9 Expert Interview.

navigating various loopholes in a weak multilateral regulatory framework purported by the OECD, has single-handedly accelerated fossil fuel production over the last decades. Moreover, a review of their policy documents suggests that G20 ECAs show absolutely no signs of addressing their past wrongdoings and effectively phasing out fossil fuels, but rather opt for allocating (minimal) finance for renewable energy projects (see 7.5) which risks facilitating an "energy addition" rather than a true energy transition (York & Bell, 2019). This is now changing but the jury is still out on how effective these changes will be.

These inadequate approaches to LFFU are arguably a manifestation of investors and financiers finding themselves being caught between a rock and a hard place. They are, after all, profit-driven nodes in the neoliberal economic web—even Foundations, despite being "non-profit" organisations, rely on the profits derived from their investment portfolios to conduct their charitable activities. Profit maximisation under various historical iterations of the capitalist mode of production has come at the expense of social and environmental devastation (Fraser, 2021), and today the tale is no different; continuously prioritising financial wealth for "beneficiaries" or "exporting economies" will inadvertently deviate from all Paris Agreement objectives (the "rock"). Conversely, effective climate action is predicated upon aggressively leaving 60–90% of all fossil fuel reserves underground (Welsby et al., 2021), which implies an inevitable and substantial "stranded asset bill" to the tune of hundreds of trillions of dollars (Linquiti & Cogswell, 2016). At face value, this bill could potentially come at the expense of pension & shareholder beneficiaries, charities and national economies, which clearly contradicts the raison d'être for these financial players (the "hard place"). A reimagined and revamped paradigm shift seems in order, but how can it be prompted?

One key may lie in redefining the raison d'être and accompanying mission statements that underpin these financial actors (see Figure 7.1). Most obviously, if Foundations pledge to take climate action through their charitable activities and programmes, they could go far beyond just bankrolling (e.g. solar PV projects) and also allocate ample funds to directly acquire and decommission existing coal, oil and gas infrastructure. A study estimates that it costs between $21,000 and $466,000 to decommission 1 MW of installed coal-fired power capacity, with an average of $117,000 per MW. Given South Africa's 38 GW of existing installed coal power capacity (see 5.3.1), this implies that decommissioning South Africa's *entire* coal fleet would cost between $800 million and $18 billion (with an average of $4.5 billion). With US Foundations alone managing some $870 billion in assets in 2014 (see 6.3),

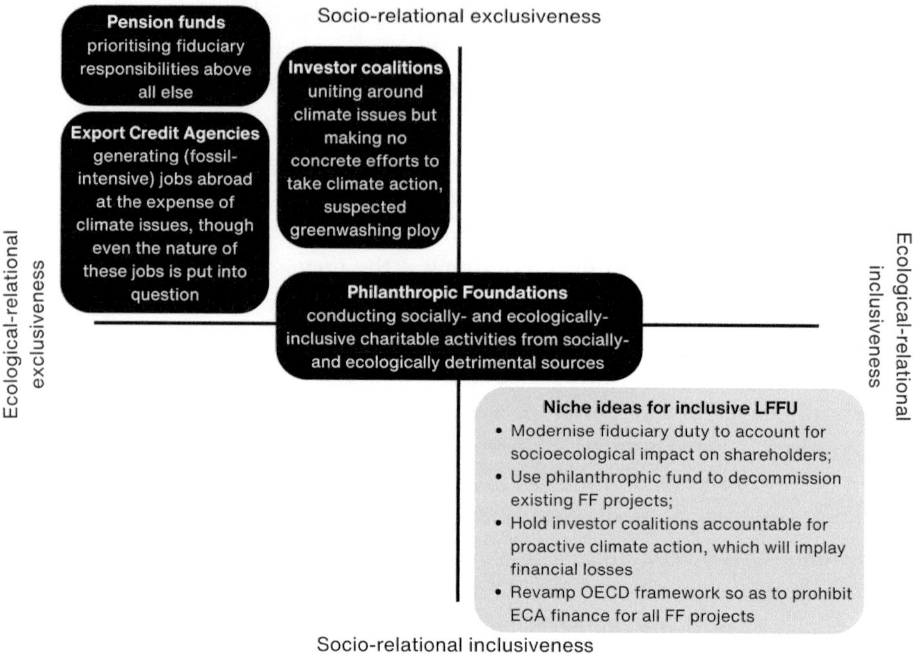

Figure 7.1. Financers and inclusive transitions.

they could *singlehandedly finance South Africa's coal phase-out 48 to 1,050 times over*. These estimates are, of course, a gross simplification and do not account for any accompanying social implications (e.g. minimising energy poverty, retraining coal miners, providing a universal basic income through a transition, etc.) or political context (see 5.3.2), but for this illustrative purpose it speaks to the order of magnitude of the funds at hand and the opportunities that are within the realm of possibility if unconventional and innovative approaches are considered.

Another key lies in the fundamental premise of relational inclusiveness. LFFU is not a win–win and utopian narrative; there will be costs, and following an inclusive approach, the most capable (and culpable) hands should bear them. For instance, the massive Dutch Pension ABP managed some €11 billion in liquid assets pertaining to fossil fuel firms in 2016 (Both Ends, 2018). Annual dividend yields for coal and oil and gas producing firms are typically greater than other sector averages and usually fluctuate between 3% and 6% (Levine, 2022), with some reaching as high as 13%. If ABP were (hypothetically) to have held onto these shares from 2016 to 2021, they would have received between €1.7 and €7.2 billion in dividends from their fossil fuel investments over five years. Drawing on the earlier example, this income entirely of fossil fuel origin could on its own finance a coal phase-out in South Africa—partially or entirely—and this is only a single Pension! Similarly, Blackrock and Vanguard—the two largest asset managers in the world and both members of the Climate Action 100+ Initiative (see 6.4)—manage $74 billion and $77 billion, respectively, in common shares pertaining to a sample of 20 major coal, oil and gas producers (as of July 2021).[10] Using current dividend yield rates for the sampled firms, Blackrock and Vanguard will jointly yield almost $12 billion ($5.8 billion each) in 2021 alone from their fossil fuel investments—also enough to partly or entirely decommission South Africa's existing coal fleet.

As such, perhaps a third element should be introduced to the divestment vs. engagement saga: a mechanism for *redistributive grants*. That is, Pensions, Foundations and asset managers that have profited handsomely from their fossil assets could forego or redistribute their earnings to directly decommission coal mines, coal-fired power stations and oil refineries, retrain workers and provide a universal income as workforces transition to different sectors and entire (fossil fuel-dependent) communities are reimagined, and cover the stranded financial assets in the form of e.g. debt that institutions have incurred to finance fossil-intensive projects (like Eskom vis-à-vis Medupi and Kusile). This proposed redistributive grant mechanism resonates with the mitigation alliance (MA) narrative in Chapter 7, and perhaps the two should be pursued in tandem to gather increased support and likelihood of success.

As alluded to earlier, this requires modernising our understanding of "fiduciary obligation" beyond simply financial interest but to encapsulate the social, ecological and relational interest of beneficiaries. The legal, political and financial implications of doing so are beyond the scope of this chapter, and we encourage future research to tackle this lacuna, but, at face value, this niche paradigm shift could spark the beginning of an effective and inclusive fossil fuel phase-out.

References

Adloff, F. (2015). Foundations and the charisma of giving: A historical sociology of philanthropy in Germany and the United States. *Voluntas: International Journal of Voluntary and Nonprofit Organizations, 26*(5), 2002–2022. https://doi.org/10.1007/s11266-014-9508-4

10 Original analysis; see methods in Rempel (2023).

Amadeo, K. (2021, March 4). Pension funds, types, top 10: Why your pension fund may be gone when you need it. *The Balance*. https://www.thebalance.com/pension-funds-definition-list- and-issues-3305875

Ameli, N., Drummond, A., Bisaro, A., Grubb, M., & Chenet, H. (2020). Climate finance and disclosure for institutional investors: Why transparency is not enough. *Climactic Change, 160*(4), 565–589.

Anabtawi, I., & Stout, L. (2008, March). Fiduciary duties for activist shareholders. *Stanford Law Review,* 60(5), 1255–1308.

Anheier, H. K., & Daly, S. (2006). Roles of foundations in Europe: A comparison. In K. Prewitt, M. Dogan, S. Heydemann & S. Toepler (Eds.), *The legitimacy of philanthropic foundations: United States and European perspectives* (pp. 192–217). Russell Sage Foundation.

Anheier, H. K., & Leat, D. (2013). Philanthropic foundations: What rationales? *Social Research, 80*(2), 449–472.

Ansar, A., Caldecott, B., & Tilbury, J. (2013). *Stranded assets and the fossil fuel divestment campaign: What does divestment mean for the valuation of fossil fuel assets?* Smith School of Enterprise and the Environment, University of Oxford. https://www.smithschool.ox.ac.uk/publications/reports/SAP-divestment-report-final.pdf

Baldwin, E., Cai, Y., & Kuralbayeva, K. (2020). To build or not to build? Capital stocks and climate policy. *Journal of Environmental Economics and Management, 100*, 102235. https://doi.org/10.1016/j.jeem.2019.05.001

BankTrack. (N.d). Kusile coal power plant, South Africa. https://www.banktrack.org/project/kusile_coal_power_plant/0/medupi_coal_power_plant

Berger, S., & Przyrembel, A. (Eds.). (2019). *Moralizing capitalism: Agents, discourses and practices of capitalism and anti-capitalism in the modern age*. Springer. https://doi.org/10.1007/978-3-030-20565-2

Bergman, N. (2018). Impacts of the fossil fuel divestment movement: Effects on finance, policy and public discourse. *Sustainability, 10*(7), 2529. https://doi.org/10.3390/su10072529

Bos, K., & Gupta, J. (2018). Climate change: The risks of stranded fossil fuel assets and resources to the developing world. *Third World Quarterly, 39*(3), 436–453. https://doi.org/10.1080/01436597.2017.1387477

Both ENDS. (2015). *Shady dealings: How Atradius Dutch State Business fails to adequately screen the business it supports*. https://www.bothends.org/uploaded_files/document/Shady_dealings_online.pdf

Both ENDS. (2018). *Still dirty, still dangerous: The fossil fuels investments of Dutch pension fund ABP.* https://www.bothends.org/uploaded_files/document/1Still_Dirty_Still_Dangerous_the_fossil_fuel_investm.pdf

Both ENDS. (2019). *The fossil elephant in the room.* https://www.bothends.org/uploaded_files/document/The_fossil_elephant_in_the_room_web.pdf

CA100+. (2020). *2020 progress report.* https://www.climateaction100.org/wp-content/uploads/2020/12/CA100-Progress-Report.pdf

Carbon Tracker Initiative. (2013). *Unburnable carbon 2013: Wasted capital and stranded assets.* Carbon Tracker and the Grantham Research Institute on Climate Change and the Environment. https://www.lse.ac.uk/granthaminstitute/wp-content/uploads/2014/02/PB-unburnable-carbon-2013-wasted-capital-stranded-assets.pdf

CPPIB [Canada Public Pension Investment Board]. (2018). *Report on sustainable investing: Investing responsibly for CPP contributors and beneficiaries.* https://www.cppib.com/documents/1922/CPPIB_SI_2018_ENG.pdf

Daly, S. (2012). Philanthropy as an essentially contested concept. *Voluntas: International Journal of Voluntary and Nonprofit Organizations, 23*(3), 535–557. https://doi.org/10.1007/s11266-011-9213-5

Damen. (2018). Small ships, tall ships, all ships. Damen Shipyards Group.

Damen. (N.d.). Offshore energy. Damen Shipyards Group. https://www.damen.com/markets/offshore-energy

McDonnell, C., A. Rempel & J. Gupta (2022). Climate action or distraction? Exploring investor initiatives and implications for unextractable fossil fuels, *Energy Research and Social Science*, 92, https://doi.org/10.1016/j.erss.2022.102769

DeAngelis, K., & Tucker, B. (2020). *Adding fuel to the fire: Export credit agencies and fossil fuel finance.* Friends of the Earth United States and Oil Change International. http://priceofoil.org/content/uploads/2020/01/2020.01.30_Adding-Fuel-to-the-Fire_final.pdf

DFCD [Dutch Fund for Climate and Development]. (N.d.). Investing for climate-resilient economic growth. https://thedfcd.com/

Erickson, P., Lazarus, M., & Piggot, G. (2018). Limiting fossil fuel production as the next big step. *Nature Climate Change, 8*, 1037–1043. https://doi.org/10.1038/s41558-018-0337-0

EXIM [Export-Import Bank of the US]. (2019). *Keeping America strong: Annual report.* https://img.exim.gov/s3fs-public/reports/annual/2019/EXIM_2019%20AnnualReport_508C_Web.pdf

Finley-Brook, M., & Holloman, E. (2016). Empowering energy justice. *International Journal of Environmental Research and Public Health, 13*(9), 926. https://doi.org/10.3390/ijerph13090926

Fisher, D. (1983). The role of philanthropic foundations in the reproduction and production of hegemony: Rockefeller Foundations and the social sciences. *Sociology, 17*(2), 206–233. https://doi.org/10.1177/0038038583017002004

Fossil Free. (2019). Global fossil fuel divestment and clean energy investment movement crosses $11 trillion milestone. https://gofossilfree.org/press-release/global-fossil-fuel-divestment-and-clean-energy-investment-movement-crosses-11-trillion-milestone/

Fossil Free UK. (N.d.). UK councils: Fuelling the fire. https://gofossilfree.org/uk/pensions/

Foundation Center. (2014). Key facts on US foundations. https://foundationcenter.org/gainknowledge/research/keyfacts2014/pdfs/Key_Facts_on_US_Foundations_2014.pdf

Fraser, N. (2021). Climates of capital for a trans-environmental eco-socialism. *New Left Review, 127*, 94–127.

Gond, J., & Piani, V. (2013). Organizing the collective action of institutional investors: Three case studies from the Principles for Responsible Investment Initiative. *Critical Studies on Corporate Responsibility, Governance and Sustainability, 5*, 19–59.

Government of Canada. (2015). *Canada's International Climate Finance.*

Grady-Benson, J., & Sarathy, B. (2016). Fossil fuel divestment in US higher education: Student-led organising for climate justice. *Local Environment: The International Journal of Justice and Sustainability, 21*(6), 661–681. https://doi.org/10.1080/13549839.2015.1009825

Green, F., & Denniss, R. (2018). Cutting with both arms of the scissors: The economic and political case for restrictive supply-side policies. *Climate Change, 150*, 73–87. https://doi.org/10.1007/s10584-018-2162-x

Gunningham, N. (2020). A quiet revolution: Central banks, financial regulators, and climate finance. *Sustainability, 12*(22), 9596. https://doi.org/10.3390/su12229596

Gupta, J., Rempel, A., & Verrest, H. (2020). Access and allocation: The role of large shareholders and investors in leaving fossil fuels underground. *International Environmental Agreements: Politics, Law and Economics, 20*(2), 303–322. https://doi.org/10.1007/s10784-020-09478-4

Hammack, D. C. (2006). American debates on the legitimacy of foundations. In K. Prewitt, M. Dogan, S. Heydemann & S. Toepler (Eds.), *Legitimacy of philanthropic foundations: United States and European perspectives* (pp. 49–98). Russell Sage Foundation.

Healy, N., & Barry, J. (2017). Politicizing energy justice and energy system transitions: Fossil fuel divestment and a "just transition". *Energy Policy, 108*, 451–459. https://doi.org/10.1016/j.enpol.2017.06.014

Henderson, R. (2020). The unlikely environmentalists: How the private sector can combat climate change. *Foreign Affairs, 99*, 47–49.

Hesketh, C. (2017). Passive revolution: A universal concept with geographical seats. *Review of International Studies, 43*(3), 389–408. https://doi.org/10.1017/S0260210517000079

Hesselmann, E. (2011). The limits of control: The accountability of foundations and partnerships in global health. In S. Rushton & O. D. Williams (Eds.), *Partnerships and foundations in global health governance* (pp. 228–252). Palgrave Macmillan.

Hinson, S., Sutherland, N., Rhones, C., Bolton, P., & Cracknell, R. (2020). Debate pack: UK oil and gas industry. House of Commons Library.

Hinz, R., Rudolph, H. P., Antolín, P., & Yermo, J. (Eds.). (2010). *Evaluating the financial performance of pension funds.* The Word Bank.

House of Commons. (2019). *UK Export Finance: Nineteenth report of session 2017–19.* https://publications.parliament.uk/pa/cm201719/cmselect/cmenvaud/1804/1804.pdf

INCITE! Women of Color against Violence. (2017). *The revolution will not be funded: Beyond the non-profit industrial complex.* Duke University Press.

IPCC [Intergovernmental Panel on Climate Change]. (2021). *IPCC sixth assessment report, Working Group 1: Climate change 2021: The physical science basis.* https://www.ipcc.ch/report/ar6/wg1/

Johnstone, P., Stirling, A., & Sovacool, B. (2017). Policy mixes for incumbency: Exploring the destructive recreation of renewable energy, shale gas "fracking," and nuclear power in the United Kingdom. *Energy Research & Social Science, 33*, 147–162. https://doi.org/10.1016/j.erss.2017.09.005

Kourula, A., & Delalieux, G. (2016). The micro-level foundations and dynamics of political corporate social responsibility: hegemony and passive revolution through civil society. *Journal of Business Ethics, 135*(4), 769–785. https://doi.org/10.1007/s10551-014-2499-7

Koushyar, J., Longhofer, W., & Roberts, P. W. (2015). A comparative analysis of corporate and independent foundations. *Sociological Science, 2*(28), 582–596. https://doi.org/10.15195/v2.a28

Kruitwagen, L., Madani, K., Caldecott, B., & Workman, M. (2017). Game theory and corporate governance: Conditions for effective stewardship of companies exposed to climate change risks. *Journal of Sustainable Finance and Investment, 7*(1), 14–36.

Lazarus, M., Erickson, P., & Tempest, P. (2015). *Supply-side climate policy: The road less taken.* Stockholm Environment Institute. https://mediamanager.sei.org/documents/Publications/Climate/SEI-WP-2015-13-Supply-side-climate-policy.pdf

Levine, S. (2022, July 1). Best oil dividend stocks of 2022. *The Motley Fool.* https://www.fool.com/investing/stock-market/market-sectors/energy/oil-stocks/oil-dividend-stocks/

Linquiti, P., & Cogswell, N. (2016). The carbon ask: Effects of climate policy on the value of fossil fuel resources and the implications for technological innovation. *Journal of Environmental Studies and Sciences, 6*(4), 662–676.

MacLeod, M. (2009). Emerging investor networks and the construction of corporate social responsibility. *Journal of Corporate Citizenship, 34*, 69–98.

Matikainen, S., & Soubeyran, E. (2022, July 27). *What are stranded assets?* Grantham Research Institute on Climate Change and the Environment. https://www.lse.ac.uk/granthaminstitute/explainers/what-are-stranded-assets/

Maurer, C. (2002). The transition from fossil to renewable energy systems: What role for export credit agencies? *OGEL Energy Law Journal, 2*. https://www.ogel.org/article.asp?key=1211

McDonnell, C., A. Rempel & J. Gupta (2022). Climate action or distraction? Exploring investor initiatives and implications for unextractable fossil fuels, *Energy Research and Social Science*, 92, https://doi.org/10.1016/j.erss.2022.102769

Mercure, J., et al. (2018, July). Macroeconomic impact of stranded fossil fuel assets. *Nature Climate Change, 8*, 588–593. https://doi.org/10.1038/s41558-018-0182-1

Morena, E. (2016). *The price of climate action: Philanthropic foundations in the international climate debate.* Palgrave Macmillan.

NZAOA [Net-Zero Asset Owner Alliance]. (2021). *Inaugural 2025 target setting protocol.* https://www.unepfi.org/publications/aoapublication/inaugural-2025-target-setting-protocol/

OECD [Organisation for Economic Co-operation and Development]. (N.d.). Funded pensions indicators. https://stats.oecd.org/Index.aspx?DatasetCode=PNNI_NEW

OECD [Organisation for Economic Co-operation and Development]. (2016). 2020 projections of climate finance towards the USD100 billion goal: Technical note. https://www.oecd.org/environment/cc/Projecting%20Climate%20Change%20 2020%20WEB.pdf

OECD [Organisation for Economic Co-operation and Development]. (2020a). *Arrangement on officially supported export credits.* https://www.oecd.org/officialdocuments/publicdisplaydocumentpdf/?doclanguage=en&cote=tad/pg(2020)1

OECD [Organisation for Economic Co-operation and Development]. (2020b). *Pension markets in focus.* https://www.oecd.org/daf/fin/private-pensions/Pension-Markets-in-Focus-2020.pdf

O'Rourke, A. (2003). A new politics of engagement: Shareholder activism for corporate social responsibility. *Business Strategy and the Environment, 12*(4), 227–239. https://doi.org/10.1002/bse.364

Partex Oil and Gas. (N.d.). The history of Partex. https://partex-oilgas.com/the-history-of-partex/

Petroleum Commission Ghana. (N.d.). *Petroleum register.* https://www.ghanapetroleumregister.com/sankofa

Rempel, A. M. (2023). *Inclusively leaving fossil fuels underground in South Africa. From a climate debt to an unsettled stranded asset debt.* PhD diss., University of Amsterdam. https://pure.uva.nl/ws/files/108548329/Thesis.pdf

Rempel, A., & Gupta, J. (2020). Conflicting commitments? Examining pension funds, fossil fuel assets and climate policy in the Organisation for Economic

Co-operation and Development (OECD). *Energy Research & Social Science, 69*, 101736. https://doi.org/10.1016/j.erss.2020.101736

Rempel, A., & Gupta, J. (2021). Fossil fuels, stranded assets and COVID-19: Imagining an inclusive & transformative recovery. *World Development, 146*, 105608. https://doi.org/10.1016/j.worlddev.2021.105608

Rempel, A., & Gupta, J. (2022). Equitable, effective, and feasible approaches for a prospective fossil fuel transition. *WIREs Climate Change, 13*(2), e756. https://doi.org/10.1002/wcc.756

Sarang, S. (2015). Combating climate change through a duty to divest. *Columbia Journal of Law and Social Problems, 49*(2), 295–341.

Schanzenbach, M. M., & Sitkoff, R. H. (2020). Reconciling fiduciary duty and social conscience: The law and economics of ESG investing by a trustee. *Stanford Law Review, 72*, 381–454.

Schneider, N. (2014). Revisiting divestment. *Hasting Law Journal, 66*(2), 589–616.

Sovacool, B. K., & Geels, F. W. (2016). Further reflections on the temporality of energy transitions: A response to critics. *Energy Research & Social Science, 22*, 232–237. https://doi.org/10.1016/j.erss.2016.08.013

Takahashi, P. (2021, June 5). Total latest oil company to rebrand as world shifts from fossil fuels. *Houston Chronicle*. https://www.houstonchronicle.com/business/energy/article/Total-latest-oil-company-to-rebrand-as-world-16226161.php

Thelin, J. R. (2014a). Government relations and the nonprofit sector: Legislation and policies in philanthropy and higher education. In J. R. Thelin & R. W. Trollinger, *Philanthropy and American higher education* (pp. 119–143). Palgrave Macmillan.

Thelin, J. R. (2014b). Philanthropists and their foundations. In J. R. Thelin & R. W. Trollinger, *Philanthropy and American higher education* (pp. 67–93). Palgrave Macmillan.

TotalEnergies. (2020, September). *Getting to net zero*. https://totalenergies.com/sites/g/files/nytnzq121/files/documents/2020-10/total-climate-report-2020.pdf

TPN/Lusa. (2019, June 17). Exploration and production, a Thai state-owned enterprise for $622 million (about $555 million). *Portugal News*. https://www.theportugalnews.com/news/gulbenkian-sells-partex-for-estimated-555m/49955

Tucker, B., & DeAngelis, K. (2020, May). *Still digging: G20 governments continue to finance the climate crisis*. Oil Change International and Friends of the Earth US. http://priceofoil.org/content/uploads/2020/05/G20-Still-Digging.pdf

UKEF [UK Export Finance]. (N.d.). About us. https://www.gov.uk/government/organisations/uk-export-finance

Ulbert, C., & Hamm, B. (2011). Private foundations as agents of development in global health: What kind of impact do they have and how to assess it? In S. Rushton & O. D. Williams (Eds.), *Partnerships and foundations in global health governance* (pp. 184–205). Palgrave Macmillan.

Van Baal, M., & Ashurst, M. (2021, January 12). Climate Action 100+ alliance is bigger than Big Oil. Responsible Investor: https://www.responsible-investor.com/articles/climate-action-100-alliance-is-bigger-than-big-oil

Vogel, A. (2010). Democratic legitimacy of philanthropic foundations: US grant-making in the Middle East. In E. Erman & A. Uhlin (Eds.), *Legitimacy beyond the state? Re-examining the democratic credentials of transnational actors* (pp. 64–84). Palgrave Macmillan.

Welsby, D., Price, J., Pye, S., & Ekins, P. (2021). Unextractable fossil fuels in a 1.5 °C world. *Nature, 597*, 230–234. https://doi.org/10.1038/s41586-021-03821-8

Williams, O. (2021, June 10). World's wealth hits half a quadrillion dollars. *Forbes*. https://www.forbes.com/sites/oliverwilliams1/2021/06/10/worlds-wealth-hits-half-a-quadrillion-dollars/?sh=26c12b18309d

York, R., & Bell, S. E. (2019). Energy transitions or additions? Why a transition from fossil fuels requires more than the growth of renewable energy. *Energy Research & Social Science, 51*, 40–43. https://doi.org/10.1016/j.erss.2019.01.008

8. Supply-side and Institutional Mechanisms for LFFU

Arthur Rempel, Carlos Larrea, Maria Rosa Murmis and Barbara Hogenboom

Abstract

How can international collaboration to effectively and inclusively leave fossil fuels underground be fostered? This chapter discusses the potential of public sector initiatives and institutional mechanisms that halt the supply of fossil fuels and the funding of fossil energy projects. Next to national policies, a transition in the role of public finance institutions is indispensable. The creation of mitigation alliances—an innovative global tool included in the Paris Agreement—offers great potential. The chapter reviews several supply-side approaches and experiences in the Global South, such as oil moratoria and the Yasuní–ITT Initiative in Latin America. After showing the important (negative) role of foreign public finance institutions in funding fossil fuel projects in countries like Ecuador and South Africa, the chapter explains how through setting up mitigation alliances they can start playing a positive role and support climate action and LFFU. A mitigation alliance can declare specific fossil fuel reserves unburnable and create compensation funds that invest in renewable energy, sustainable social development and conservation of forests and biodiversity in developing countries.

Keywords: climate action, public finance institutions, international collaboration, mitigation alliances, Global South

8.1. Introduction

The world is not on track to fulfil the goals of the Paris Agreement (PA) (UNFCCC, 2015). Current pledges, if fulfilled, will lead to a global temperature increase

Gupta, J., Hogenboom, B., Rempel, A. & Olofsson, M. (eds), *Leaving Fossil Fuels Underground: Actors, Arguments and Approaches in the Global South and Global North.*
Taylor & Francis Group, 2024
doi: 10.5117/9789048560370_CH08

of 2.5 °C by 2100, well beyond the Paris limit of 1.5–2 °C. Current mitigation policies have been insufficient, and a deep change is required. GHG emissions must be cut by 45% in 2030, and most countries must become carbon neutral by mid-century to avoid the catastrophic effects of climate change (UNEP, 2022). Furthermore, according to scientific evidence, at least two-thirds of fossil fuel reserves must remain in the ground indefinitely in order to fulfil the PA goals (McGlade & Ekins, 2015; Meinshausen et al., 2009; IEA, 2012; Welsby et al., 2021). Given the limited success of climate policy so far, policy instruments targeting the supply of fossil fuels deserve far more attention. Until recently, mitigation policies have been strongly focused on demand restrictions. The main mechanisms implemented include cap-and-trade, carbon taxes and policies for promoting R&D and implementation of renewable energy and improving energy efficiency (see Rempel & Gupta, 2022). The results of prevailing cap-and-trade systems have been limited, mostly due to low carbon prices and limited coverage (Sinn, 2012; Armstrong, 2019). In 2022, only 24% of world CO_2 emissions were being regulated by carbon prices, and despite recent carbon price increases, less than 4% of emissions were covered by carbon prices above the level required to meet the 2030 goals (World Bank, 2022).

Comparatively little (but growing) attention has been paid to mitigation via fossil fuel supply restrictions (cf. Lazarus et al., 2015; Green & Denniss, 2018; Le Billon & Kristoffersen, 2019; Gaulin & Le Billon, 2020; Rempel & Gupta, 2022). In addition, approaches that account for the massive international financial flows that drive global fossil fuel development have been largely omitted from the academic and policy space. In particular, supply-side policy (proposals) in the Global South, and a shift in global funding that can support such policies, need to be further examined and enhanced. In this respect, the transition in global funding includes a key role for public finance institutions (PFIs)—that is, multilateral and bilateral development banks (MDBs and BDBs, respectively) and export credit agencies (ECAs). The PFIs from G20 states were found to have allocated almost $1 trillion to fossil fuel projects globally through loans, guarantees, debt relief and underwritings from 2007 to 2019 (see Rempel & Gupta, 2022).

This chapter aims to answer the question: How can international collaboration be fostered to effectively and inclusively leave fossil fuels underground? We are particularly interested in public sector initiatives and mechanisms that halt the supply of fossil fuels and the funding of fossil energy projects. Next to national policies this includes a transition in the role of PFIs, for which we discuss the possibilities of mitigation alliances (MAs). We first present the chapter's connection to the analytical framework of this volume (see 8.2) and elaborate on supply-side approaches and experiences in the Global South, such as oil moratoria and bans and the Yasuní–ITT Initiative in Latin America (see 8.3). We then explore the international financial linkages between public finance institutions and Latin American and African fossil fuel projects (see 8.4), propose MAs as an innovative global approach to LFFU building on the PFI finance flows (see 8.5), and finally draw conclusions (see 8.6).

8.2. Building on the analytical framework

This chapter focuses on approaches and arguments that are relatively understudied and overlooked in international mitigation policies. The urgent need for more effective mitigation results has led to increasing support for the idea of complementing existing demand-based policies with new mechanisms to limit fossil fuel supply (Lazarus et al., 2015; Green & Denniss, 2018). Supply-side measures include, inter alia: removal of monetary and non-monetary funding and subsidies that incentivise fossil fuel exploration and production (e.g. Johnsson et al., 2019; Coady et al., 2019; Erickson et al., 2018; Lazarus & Van Asselt, 2018); moratoria and bans that temporarily (moratoria) or permanently (bans) prohibit fossil fuel resources from a particular reserve or using a particular technology (e.g. unconventional gas fracking) (cf. Le Billon & Kristoffersen, 2019; Lazarus & Van Asselt, 2018); compensatory mechanisms, or "finance swaps," in which a nation agrees to forgo exploiting its coal, oil or gas reserves in exchange for financial compensation from one or multiple external parties (e.g. Martin & Scholz, 2014; Kingsbury et al., 2018; Le Billon & Kristoffersen, 2019); and additional institutional mechanisms to determine the reserves to remain untapped, such as an auction mechanism (Pellegrini et al., 2021). These and other supply-side approaches to restrict the supply and thereby the extraction of fossil fuels are an indispensable part of the climate agenda and global energy transition. Supply-side restriction or mitigation avoidance policies have significant advantages over conventional mitigation approaches when implemented in tandem with existing demand-restricting policies. They have been regarded as: (a) more cost-effective, (b) requiring lower transaction costs, (c) having higher verification and monitoring efficiency than conventional mechanisms, and, under certain conditions, (d) less prone to leakage (see e.g. Lazarus & Van Asselt, 2018; Gaulin & Le Billon, 2020; Sinn, 2012). Additionally, supply restriction policies will provide an effective tool to prevent the so-called "green paradox," consisting of an acceleration of extraction caused by current fossil fuel producers attempting to convert their wealth in the ground into cash faced with prospects of future demand restrictions, due to more stringent environmental regulation and lower prices (Sinn, 2012).

The key actors in all of the abovementioned initiatives to restrict the fossil fuel supply are public institutions at the national and/or multilateral level. In countries of the Global South, national governments and other public institutions have limited capacity to do so without external support or compensation, but in some cases social movements, intellectuals and politicians have successfully mobilised ideas for not exploiting fossil fuel reserves. Nevertheless, due to foregoing some short-term economic and political advantages, and without institutionalised external funding, laudable policy initiatives to leave fossil reserves untouched can easily be withdrawn by a shift in government and policy (see 8.3). Especially foreign public finance institutions—ranging from bilateral and multilateral development banks to export credit agencies—are crucial actors in the success or failure of the energy transition of the Global South. These PFIs are public institutions; their main shareholders are either one (in the case of BDBs and ECAs) or multiple (in the case of MDBs) governments which allocate public funds for domestic and international projects. In the case of development banks, these funds (usually in the form of loans or underwritings) are typically stipulated under the premise that they serve the "development" of the recipient (i.e. indebted) nation. ECAs operate slightly differently in that they often play a crucial role in insuring (i.e. guaranteeing) a potential overseas project for a domestic company or various firms (Gupta et al., 2020; Rempel & Gupta, 2022). As we will see in 8.4, even since the Paris Agreement and its explicit limits to such arrangements, PFIs continue to massively support fossil fuel extraction in the Global South, de-risking such projects which can then attract supplemental private funding. Based on an original empirical analysis of the public finance flows to Ecuador and South Africa (using the expansive Shift the Subsidies Database of Oil Change International), we clearly see that PFIs are lagging behind in the energy transition. The approach of voluntary MAs discussed in 8.5 proposes a radically different and positive role of PFIs, in which they can support effective supply-side measures in the Global South that are socially, environmentally and relationally inclusive, in line with the inclusive development perspective discussed in 2.2.3. In addition to more effective climate mitigation, the comprehensive nature of this proposal can be found in eradicating fossil fuel extraction in vulnerable areas, funding clean energy and addressing issues of socio-environmental and restorative justice.

In short, with an analysis of supply-side and institutional mechanisms for LFFU, a presentation of data on the PFIs' currently failing support for this in the Global South, and an innovative proposal for MAs, this chapter looks into actors, arguments and approaches for LFFU that complement those analysed in Chapters 3, 4, 5 and 6. In addition, this chapter identifies various niche examples of supply-side LFFU (8.3), the ample room for improvement in the agency of PFIs (8.4), and a concrete institutional mechanism for the energy transition with inclusive development and socio-environmental justice (8.5). Identifying the bottlenecks and possibilities of scaling up fossil fuel exploitation and exploration bans and implementing MAs will mean an important step in the inclusive energy transition.

8.3. Supply-side approaches: From national bans to international compensation

Knowing that a large portion of fossil fuel reserves must be left underground (Welsby et al., 2021), effective supply-side restrictions must consider which reserves may be exploited and which must remain underground. The conventional criterion for selection is based on exploiting reserves with the lowest extraction costs, at best with restrictions linked with energy carbon content (McGlade & Ekins, 2015). The extraction cost criterion traditionally does not include the negative environmental repercussions resulting from fossil fuel extraction, which are particularly high in biodiversity hotspots or other sites of high conservation value, including cultural and social value. Biodiversity loss and the ensuing deterioration of local and global ecosystem services is one of the most serious global environmental problems faced today (Steffen et al., 2015), together with climate change. In various supply-side mechanisms for LFFU, the agendas for halting climate change and preventing biodiversity loss are combined. We see this in the case of the Latin American moratoria and bans in Section 8.3.1 and this was also the case for the only compensatory mechanism attempted to date: the Yasuní–ITT Initiative of Ecuador, which we look into in 8.3.2. Although it was short-lived (2007–2013), the implementation of the Yasuní–ITT Initiative offers numerous lessons that can be adopted to devise a new, more progressive and effective supply-side approach to LFFU in the global arena.

8.3.1. Latin American oil moratoria and bans

Despite the increasing interest in curtailing the extraction and supply of fossil fuels, studies on such policies are scarce and primarily discuss their potential pros and cons (cf. Erickson et al., 2018; Green & Denniss, 2018; Le

Billon & Kristoffersen, 2019; Lazarus & Van Asselt, 2018), as there are only a few cases where they have already been implemented. Latin America is, however, a relative frontrunner in the field of these new "supply-side" policies for LFFU (Tudela, 2020). Next to a ban on oil and gas activity in some of Mexico's biosphere reserves since 2014, two national oil drilling moratoria have been installed in Central America: in Costa Rica and Belize. In 2002 Costa Rica initiated its first moratorium on oil exploration, and between 2011 and 2021 oil exploitation was temporarily banned via moratorium policies of the presidents of Chinchilla and Solís. In December 2021, the government announced a plan to implement a law that declares Costa Rica's national territory free of oil and gas exploration and exploitation. In Belize a moratorium on all oil-related activities in maritime areas was adopted in 2017 to protect the maritime environment, including its reefs recognised by the United Nations Educational, Scientific and Cultural Organization (UNESCO) as World Heritage. In both cases, environmental groups were the main drivers behind the moratoria, by initiating and mobilising people for the idea (see Chapter 4) and initiating and winning court cases (see Chapter 3) to overcome political or legal obstacles. At the same time, we notice that the idea of a moratorium became influential when it was bundled with a multi-actor coalition in which, for instance, the tourist industry also supported banning oil extraction (similar developments can be found in recent initiatives for mining moratoria, cf. Spalding, 2018). Tudela (2020) shows that protection of biodiversity and ecosystem services—rather than a climate rationale—provided the imperative for the oil moratoria and bans in Belize, Costa Rica and Mexico. Protecting rich biodiversity, natural heritage, associated cultural values and the environment at large were the main reasons, next to protecting the more economic interests of the ecotourism and fishery sectors.

While someone might dismiss oil drilling moratoria in small nations with limited reserves such as Belize and Costa Rica as being of limited relevance, they point at the room for ideational, political and institutional innovation and supply-side LFFU policies, which can be scaled up. A moratorium on fossil fuel exploration and extraction is still an underdeveloped and underused legislative tool for LFFU, implementing a profoundly different attitude towards fossil fuels. It prevents all activities to explore and exploit the reserves of a particular resource within a certain territory. To use such a ban in other settings than highly protected areas reflects consciousness of the ecological, social and economic downsides and the risks of exploiting fossil fuels, up to the point that its commodification is no longer perceived as beneficial or acceptable, despite the loss of certain revenues. Still, to leave fossil fuel reserves completely untouched requires overcoming powerful economic, political and social interests and logics served by exploitation, and to break with the extractive imperative that dominates in policies on non-renewables in Latin America and beyond (Arsel et al., 2016). Therefore, in order to scale up bans on fossil fuel exploitation to larger countries and other regions we also need to review proposals to embed them in international arrangements that provide institutional backing and a fair compensation to the society of a developing country. An emphasis on all the short-term and long-term potential benefits to be reaped from bans and moratoria, including reaching the climate goals of the Paris Agreement, may increase the viability of their adoption.

8.3.2. Ecuador's Yasuní–ITT Initiative

Ecuador's Yasuní–ITT Initiative (2007–2013) was the first, and is still the only, international proposal presented by a developing oil-exporting country (see Chapter 4) to keep a large oil reserve indefinitely unexploited, in exchange for an international fund to be invested only in forest conservation, clean energy transition and inclusive social development. In 2007 a large oil reserve—850 million barrels of oil—was confirmed in the ITT oil field, located in the Yasuní National Park. The park has been regarded as one of the most biodiverse hotspots in the Western Hemisphere (Bass et al., 2010) and it is home to the only two Indigenous groups living in voluntary isolation in Ecuador (Tagaere and Taromenani). In September 2007 Ecuador presented at the United Nations the Yasuní–ITT Initiative. Ecuador committed itself to keep the ITT reserve indefinitely unexploited, in exchange for an international compensation fund equivalent to at least half of the expected government earnings in case of oil extraction. The fund should be invested only in renewable energy, increasing energy efficiency, forest conservation and sustainable social development in the Amazon. In 2010 an international agreement was signed between Ecuador and the UNDP, creating the fund under UN supervision, to guarantee transparency and efficiency (for a detailed overview, see Larrea et al., 2009; Larrea & Warnars, 2009; Vallejo et al., 2015).

The project has received the official support of various internationally recognised individuals, including; Nobel Peace Prize laureates Muhammad Yunus, Desmond Tutu, Jody Williams and Rigoberta Menchú; Rita Levi Montalcini, Nobel laureate in Medicine; Ban Ki-Moon, secretary-general of the United Nations; ex-presidents Mikhail Gorbachev (USSR), Felipe González (Spain), Fernando Henrique Cardoso (Brazil), Ricardo Lagos (Chile); Prince Charles of Great Britain and Danielle Mitterrand, president of the

France Libertés Foundation, among others. It also received formal backing from the German parliament, with unanimous support from all political parties, as well as the European Union, and other international bodies such as OPEC (Organization of the Petroleum Exporting Countries), CAN (Andean Community of Nations), CAF (Andean Development Corporation), the Organization of American States, numerous international organisations, like the IUCN (International Union for Conservation of Nature), and various Indigenous organisations and ecological groups in Ecuador.

The contributions were expected to come from national governments, international institutions, regional governments, cities, civil society and individuals. In exchange, Ecuador would issue Yasuní Guarantee Certificates, a non-tradable document. Therefore, contributions were voluntary and were not part of carbon markets. Chile became the first country in the world to contribute financially to the initiative, followed by Spain. Other countries, such as Italy, Colombia, Georgia, Turkey and Peru also contributed to the fund, as well as the Wallonia regional government in Belgium and several local governments in France. Non-governmental organisations, such as AVINA, also participated in the initiative.

Despite significant international support, President Correa cancelled the initiative in 2013, arguing that the received funds were insufficient. According to critical evaluations, international support was adequate, but the cancelation was mostly due to lack of consistent and effective support from the Ecuadorian government, commitments to increase oil extraction in exchange for Chinese loans and the inconsistent support from President Correa, who publicly discouraged donations, removed several of the initiative's managers and persistently threatened to extract oil from the ITT fields (Larrea et al., 2021; Martin & Scholz, 2014). After cancelling the Initiative, the government started oil extraction in 2016, in association with the Chinese company Sinopec. The cancellation unleashed social opposition and a strong social movement emerged (Yasunidos, see 4.4.2). More than 700,000 signatures were collected in 2014, asking for a national referendum about oil exploitation in the ITT field. While the Constitution of Ecuador recognises the right of citizens to call a national referendum, the petition was denied in what seems to have been a fraudulent decision by President Correa, but recently Ecuador's Constitutional Court accepted it, and a future national referendum may be held.

Despite its cancellation, the initiative provided a framework to foster the transition away from fossil fuels in oil-exporting megadiverse developing countries. It was designed as a replicable tool and many of its key components are also used in the MAs that we propose in 8.5. As at least two-thirds of fossil fuel reserves must be kept unexploited to fulfil PA goals, oil-exporting developing countries need international support to foster an adequate transition towards a sustainable economy based on clean energy. In the case of Ecuador, the Yasuní–ITT Initiative model and fund was designed as a tool to foster a smooth transition towards a sustainable and equitable economy, while leaving the fossil fuel reserves located below a biodiversity hotspot untouched. As a significant fraction of fossil fuel reserves are located in vulnerable areas in developing countries, and the environmental impact of extracting them is high, an international fund to allow countries to keep those reserves unexploited may protect biodiversity, reduce emissions from both unexploited fossil fuels and avoided deforestation, and foster the transition towards low-emission sustainable development paths in these countries.

8.4. Exposing the finance flows

As mentioned in 8.2, PFIs are public institutions that have played (and continue to play) a crucial role in driving fossil fuel projects around the globe, especially by means of de-risking, which allows these projects to attract commercial finance. While MDBs and BDBs de-risk by bringing in capital, ECAs de-risk through their role as guarantor. For example, South Africa's Medupi coal-fired power plant was not only heavily financed by the World Bank (via a $3.75 billion loan issued by the IBRD and IDA) and the African Development Bank ($2.3 billion loan), but it also saw German and French ECAs (KfW IPEX and COFACE, respectively) guarantee the syndicated loans (worth $740 million and $1.2 billion, respectively) issued by their respective nation's commercial banks (Commerzbank, BNP Paribas and Credit Agricole, among others) in addition to the German and French companies (Hitachi Power Europe, Siemens, Alstom, Evonik) that were contracted to build the power plant's turbines and boiler units (see BankTrack, n.d.).

This section explores the extent to which some of the world's largest PFIs have financed energy projects in both Ecuador (see 8.4.1) and South Africa (see 8.4.2) both before and after the Paris Agreement was signed, comparing the PFI finance flows of the period from 2007 to 2015 with those of the period from 2016 to 2019. The governments of these financial institutions have all agreed to uphold the various commitments made in Paris. The Paris Agreement on Climate Change is awash with pledges that explicitly shed light on the critical role that global finance flows play in meeting the 1.5–2 °C goal (Article 2.1a). Particularly, the third objective of the agreement (Article 2.1c) concerns aligning finance flows with climate change mitigation and calls on states to "mak[e] finance flows consistent with a pathway towards low greenhouse gas emissions and climate-resilient development." Arguably, one cannot expect PFIs to immediately be able to fully redirect their policies and large finance portfolios, but a comparison of these two periods can at least provide an indication of whether PFIs are on the right track of meeting the PA commitments. The empirical analysis of these public finance flows draws on the expansive Shift the Subsidies Database, which Oil Change International shared with us (OCI, 2020). The dataset consists of 12,334 unique financial flows between 148 PFIs (34 ECAs, 100 BDBs and 14 MDBs) and global energy projects from 2007 to 2019, including new coal, oil and gas projects and investments in improving existing fossil fuel infrastructure. A few limitations merit attention. First, the period from 2008 to 2012 was characterised by the global financial crisis, which also affected the finance flows of PFIs. Second, the "pre-Paris" period spanning from 2007 to 2015 covers nine years, whereas the "post-Paris" period covers merely four, meaning that aggregate lending in the two periods is not comparable. As such, we present both aggregate lending in addition to annual lending in both periods to better unpack PFI lending to fossil fuels in the broader context of recent global climate negotiations. Finally, due to some limitations, the data should be interpreted as representative of a lower-end estimate of PFI-driven energy financing.[1]

8.4.1. Ecuador

The sampled PFIs allocated at least $725 million for fossil fuel projects in Ecuador from 2016 to 2019 (since the Paris Agreement) (see Table 8.2),

1 A brief note on the data and methodology in this section. Oil Change International shared the dataset with us via email on 30 November 2020. The dataset consists of financial flows between PFIs from both Annex 1/B and Non-Annex 1/B nations and global energy projects. The comprehensive dataset has metrics, including financial mechanisms that the PFI utilised (e.g. loan, guarantee, grant, debt relief, risk management); amount financed in original currency; amount in USD; financing institution; and project details. As Oil Change International denoted, the dataset has some limitations. Entries prior to 2012 are not comprehensive except for those pertaining to major MDBs, and some entries from 2019 were still being updated given the lagged nature of reporting, so findings regarding 2006–2012 and 2019 likely represent lower limits. Second, some PFIs have not transparently disclosed the breadth of their energy funding, particularly from countries like South Africa, again implying that these data should be interpreted as representative of a lower-end estimate of PFI-driven energy financing. We scrutinised the dataset to identify noticeable trends in PFI lending for fossil fuel energy projects globally before and after the Paris Agreement was adopted. These trends were explored through descriptive and visualisation techniques by writing a script to group the dataset elements into regional, temporal and actor-based subsets by utilising Python's Pandas package.

with an average of $181 million annually. Comparatively, before the Paris Agreement (from 2007 to 2015) fossil finance stood at $1 billion (see Table 8.1), or $111 million annually on average. Bilateral development banks (BDBs) spearheaded Ecuador-bound fossil financing before the Paris Agreement ($940 million, or 94%), while they tied with ECAs for post-Paris fossil finance ($323 million each, or 45%). These increased fossil flows coincided with more humble financial pools allocated for clean energy projects before the Paris Agreement ($259 million gross, less than $29 million annually) and after ($586 million gross, less than $150 million annually). Energy projects classified as "other" include non-solar and non-wind renewables (e.g. hydropower) as well as projects that are ambiguously disclosed and may very well entail fossil elements.

Table 8.1. Finance for Ecuadorian energy projects, 2007–2015 (in million USD)

	Fossil Fuel	Clean	Other	Total
Bilateral	$940.00	$7.00	$0.00	$947.00
Multilateral	$60.50	$1.56	$1,108.65	$1,170.71
Export Credit	$0.00	$250.00	$509.00	$759.00
Total	$1,000.50	$258.56	$1,617.65	$2,876.71

Source: Original, using data from OCI (2020).

Table 8.2. Finance for Ecuadorian energy projects, 2016–2019 (in million USD)

	Fossil Fuel	Clean	Other	Total
Bilateral	$323.33	$80.00	$0.00	$403.33
Multilateral	$78.40	$480.57	$588.40	$1,147.37
Export Credit	$323.33	$25.00	$25.00	$373.33
Total	$725.06	$585.57	$613.40	$1,924.03

Source: Original, using data from OCI (2020).

Table 8.3 and Table 8.4 isolate the top fossil fuel financiers by both type and nationality. Notably, Chinese BDBs were responsible for the bulk of fossil fuel finance before 2016 ($800 million, or 80%), and jointly with Chinese ECAs were responsible for the post-Paris finance ($323 billion, or 45% each). All of this Chinese BDB finance, which took the form of two loans to Ecuador's Ministry of Finance in 2010 and 2011 with a fixed 6–7% interest rate and 4- to 8-year maturity period, was used to develop Ecuador's oil sector, some of which would be "repaid by the sale of oil or fuel oil from the state-run PetroEcuador to China's state-owned PetroChina" (OCI, 2020). Moreover, this also included a more humble $2.5 million loan from the Inter-American Development Bank to Ecuador's state-owned oil and gas company to supply gas-powered electricity. Conversely, in 2016 the Chinese BDB and ECA finance post-Paris also raised a $646 million loan to PetroEcuador "under difficult market conditions due to low oil prices"; this loan established a "five-year crude oil sale and purchase contract" between PetroEcuador and PetroChina International.

Table 8.3. Top financiers of Ecuadorian energy projects, 2007–2015 (in million USD)

Country	Type Financier	Fossil Fuel	Clean	Other	Total Energy
China	Bilateral	$800.00	$0.00	$0.00	$800.00
Russian Federation	Bilateral	$140.00	$0.00	$0.00	$140.00
Inter-American Development Bank	Multilateral	$60.50	$1.56	$1,106.15	$1,168.21
China	Export Credit	$0.00	$250.00	$509.00	$759.00

Source: Original, using data from OCI (2020).

Table 8.4. Top financiers of Ecuadorian energy projects, 2016–2019 (in millions USD)

Country	Type Financier	Fossil Fuel	Clean	Other	Total Energy
China	Bilateral	$323.33	$0.00	$0.00	$323.33
China	Export Credit	$323.33	$0.00	$0.00	$323.33
Inter-American Development Bank	Multilateral	$78.40	$348.57	$438.40	$865.37

Source: Authors' calculation, using data from OCI (2020).

8.4.2. South Africa

Table 8.5 and Table 8.6 illustrate the finance flows linking global PFIs to South African fossil fuel projects, again both before and after the Paris Agreement was ratified. Before the Paris Agreement, the discrepancy between fossil and non-fossil finance flows stands out: the PFIs allocated over $11 billion for fossil fuel projects in the time period ($1.3 billion annually), while allocating less than $6 billion for clean energy ($650 million annually) and slightly over $1.8 billion for "other" energy projects ($200 million annually).

In this period from 2007 to 2015, almost 60% of all PFI financing for South African energy projects were used to develop fossil fuel projects, directly exacerbating the climate emergency and deepening Africa's carbon lock-in.

Since the Paris Agreement was ratified, from 2016 to 2019 over $4.4 billion ($1.1 billion annually) was allocated to developing the South African fossil fuel industry, while a more humble $1.9 billion ($450 annually) was allocated to developing the continent's solar PV and wind power capacities (see Table 8.5). Furthermore, $1.8 billion ($460 annually) was used to finance "other" energy projects, which again, include water power infrastructure in addition to ambiguously disclosed projects. Overall, since 2016, almost 55% of the sampled PFI finance for South African energy projects was funnelled to the sector of fossil fuels, which, like in the Ecuadorian case, are in direct misalignment with the goals set forth in the PA (Articles 2.1a and 2.1c).

Table 8.5. Total finance for South African energy projects, 2007–2015 (in millions USD)

	Fossil Fuel	Clean	Other	Total Energy
Bilateral	$627.78	$1,901.98	$761.92	$3,291.69
Multilateral	$5,689.89	$1,336.89	$564.67	$7,591.46
Export Credit	$5,163.48	$2,642.14	$505.30	$8,310.92
Total	$11,481.16	$5,881.01	$1,831.89	$19,194.06

Source: Original, using data from OCI (2020).

Table 8.6. Total finance for South African energy projects, 2016–2019 (in millions USD)

	Fossil Fuel	Clean	Other	Total Energy
Bilateral	$4,403.33	$679.22	$819.91	$5,902.47
Multilateral	$0.00	$1,021.70	$941.91	$1,963.61
Export Credit	$78.49	$263.02	$0.29	$341.79
Total	$4,481.82	$1,963.94	$1,762.11	$8,207.88

Source: Original, using data from OCI (2020).

Table 8.7 and Table 8.8 shed light on the geographical origins of the South Africa-bound and fossil-driving finance flows, both before and after the Paris Agreement's ratification. Since the Paris Agreement, Chinese BDBs (specifically the China Development Bank) have spearheaded PFI finance for African fossil fuels by allocating over $4.3 billion. These flows were composed of two loans in 2016–2017 aggregating to $1.83 billion to Eskom, South Africa's state-owned power utility to finance Medupi, South Africa's largest and most controversial 4,800 MW coal-fired power station originally funded by the World Bank (among other European and Asian PFIs) in 2008–2009, and a $2.5 billion loan in 2018 to finance Kusile, Medupi's sister 4,800 MW coal-fired station. Post-Paris fossil support was also notable from Export Development Canada, which issued a $78 million loan to Sasol (South Africa's home-grown coal-to-liquids pioneering oil conglomerate) for "working capital and general corporate purposes" (OCI, 2020), and the Overseas Private Investment Corporation (OPIC, US-based BDB), which issued a $40 million guarantee to finance and construct a 52 km natural gas pipeline spearheaded by a US firm, Tetra4 Propriety Limited.

Table 8.7. Top financiers of South African energy projects, 2007–2015 (in million USD)

Country	Type Financier	Fossil Fuel	Clean	Other	Total Energy
World Bank Group	Multilateral	$3,040.00	$565.39	$60.30	$3,665.69
African Development Bank	Multilateral	$2,649.89	$325.24	$382.17	$3,357.31
France	Export Credit	$2,353.45	$0.00	$0.00	$2,353.45
Germany	Export Credit	$1,484.45	$89.56	$0.00	$1,574.01
United States	Export Credit	$822.12	$22.58	$0.00	$844.70
Japan	Export Credit	$503.46	$2,530.00	$0.00	$3,033.46
Germany	Bilateral	$444.30	$191.81	$324.72	$960.84
South Africa	Bilateral	$167.03	$968.81	$0.00	$1,135.84
Netherlands	Bilateral	$16.45	$20.00	$0.00	$36.45

Source: Original, using data from OCI (2020).

Table 8.8. Top financiers of African energy projects, 2016–2019 (in million USD)

Country	Type Financier	Fossil Fuel	Clean	Other	Total Energy
China	Bilateral	$4,333.33	$0.00	$273.89	$4,607.22
Canada	Export Credit	$78.45	$0.00	$0.00	$78.45
United States	Bilateral	$40.00	$0.00	$7.08	$47.08
Germany	Bilateral	$30.00	$176.14	$107.22	$313.36
United Kingdom	Export Credit	$0.04	$104.88	$0.29	$105.20

Source: Original, using data from OCI (2020).

PFI finance for African fossil fuels from 2007 to 2015 was even greater than that post-Paris. This was largely driven by multilateral finance ($5.7 billion, or 49% of total), including two gigantic loans in 2009 by the World Bank's International Bank for Reconstruction and Development (IBRD) ($3.1 billion) and the African Development Bank ($2.65 billion), both of which were used to finance the aforementioned Medupi coal power station. Note that French and German ECA finance ($2.4 billion and $1.5 billion, respectively) was also used to catalyse and drive the Medupi and Kusile power station projects; these flows themselves attracted a slew of commercial and private finance that would otherwise not have arisen had the PFIs not interjected (Rempel, 2023).

8.5. Mitigation alliances for compensation

Given their prominent role in fossil fuel finance, PFIs and the governments that act as their major shareholders are in an optimal position to partake in a mitigation alliance to promote effective and inclusive supply-side measures to LFFU. After all, these are *public* institutions, and their governments have all agreed to uphold the various commitments made in Paris in 2015, including, aligning finance flows with climate change mitigation (Article 2.1c), and providing monetary and non-monetary support to developing countries to improve their mitigation prospects (Article 9). Article 4.5 notes that "[s]upport shall be provided to developing country Parties ... allow[ing] for higher ambitions in their actions," through which "[d]eveloped country Parties shall provide financial resources to assist developing country Parties with respect to both mitigation and adaptation" (Article 9.1), and will also "take the lead in mobilising climate finance from a variety of sources, instruments and channels ... represent[ing] a progression beyond previous efforts" (Article 9.3, see also other applicable clauses distributed in Articles 9–11). Moreover, the Paris Agreement embodies a voluntary component through Article 6, which notes that "[p]arties recognise that some Parties choose to pursue voluntary cooperation in the implementation of their nationally determined contributions (NDCs) to allow for higher ambition in their mitigation and adaptation actions" (Article 6.1), and that the "use of internationally transferred mitigation outcomes to achieve nationally determined contributions under this Agreement shall be voluntary and authorised by participating Parties" (Article 6.3).

Whether through a voluntary (Article 6) or arguably more involuntary basis (Articles 2.1c, 4.5, 9–11), we see a tremendous opportunity to devise an innovative, supply-side mechanism for inclusively LFFU. We propose a specific avoidance mechanism that strives to keep fossil fuel reserves unexploited in megadiverse regions in developing countries, using economic compensatory measures, including both conventional and alternative instruments. The mechanism may be implemented by a voluntary mitigation alliance. Article 6 of the Paris Agreement (PA) allows for such voluntary MA of subsets of countries, financial institutions and/or subnational entities (regions, cities, private groups, etc.). Such MAs may be an important tool for LFFU in fossil fuel reserves located below megadiverse hotspots, which happens to be the case in no less than one-third of current extraction projects. A study by the World Wildlife Fund and others finds extractive industry concessions or activity in 31% of Natural World Heritage Sites: Africa, 61%; Asia- Pacific, 34%; Latin America and the Caribbean, 31%; Arab States, 17%; and Europe and North America, 10%. Significant portions of fossil fuel reserves are in megadiverse developing countries: 22.5% of oil, 8.7% of natural gas and 31.3% of coal reserves; and for all 68 biodiversity-hotspot developing countries these figures increase by 3.6, 3.8 and 0.2 percentage points, respectively (Aviva et al., 2015).

The MA must set up a common goal, an independent certification mechanism and a distribution commitment rule among its members. MAs may contribute to closing the gap between PA goals and commitments, addressing one of the main limitations of the PA (Newell & Simms, 2020; Stua, 2017). The MA can declare unburnable fossil fuel reserves, defining a certification mechanism and a valuation criterion. The compensation funds can be transferred to developing countries and be earmarked for investments in renewable energy, biodiversity conservation, avoided deforestation and sustainable social development, following the model of the Yasuní–ITT Initiative (but not limited to it). A multilateral monitoring and evaluation body must be defined. We suggest an approach involving a comprehensive concept of "cost" that takes also into account the potential loss of conservation value of the areas lying on top of the deposits in the process of determining the fossil fuel reserves that are to be left in the ground. As complex problems implying consequences in different realms (environmental, social, cultural and economic) cannot be reduced into a single monetary dimension, and it is not possible to avoid intrinsic uncertainty at valuing biodiversity loss, multicriteria analysis is an adequate tool for (a) selecting appropriate "leave in the ground reserve" options located in biodiverse hotspots and (b) ranking the best options for providing economic and equivalent compensations to allow certain reserves to remain unexploited (Munda, 2008).

Relevant remaining issues include the definition of the valuation criteria for defining compensation, permanence, leakage, sources of funding and the structure of compensation. To define the economic value of the compensation, the social cost of carbon (SCC) can be applied, and the amount of compensation must be comparable to the foregone state revenues from fossil fuel extraction, to make the conservation alternative feasible. Compensation may include not only a given economic value, but also a set of incentives and mechanisms leading to the adoption of a low emission path for development in receiving countries ("equivalent compensation measures").

An effective mechanism for keeping fossil fuel reserves underground in megadiverse or culturally sensitive hotspots in developing countries will address simultaneously the goals of climate change mitigation, biodiversity conservation and participatory social development and respect for local livelihoods. This synergic effect is also an effective way to cutting emissions, strengthening the natural resilience of carbon sinks, such as tropical rainforests, and improving the lives of the local populations. Avoided emission will then come not only from unburnable fossil fuel reserves, but also from prevented deforestation, resulting from both nature conservation and social and cultural synergies. In the example of the Yasuní–ITT Initiative, avoided emissions from unextracted oil would have reached at least 407 million tonnes of CO_2, and additional benefits from avoided deforestation were expected to reach 800 million tonnes of CO_2 (Larrea & Warnars, 2009).

Figure 8.1 provides an overview of how an MA may operate. First, an alliance of financial and political organisations is formed, including those that are either *capable* of providing supply-side mitigation support as is established in the Paris Agreement, or *culpable* for previous fossil fuel support and de facto misalignment with the Paris Agreement and broader climate objectives; the latter point draws on a *restorative justice* argument (Le Billon & Kristoffersen, 2019), namely that nations and institutions that have enjoyed benefits from commercialising, extracting and combusting fossil fuels—and have thus accumulated a *climate debt* (see e.g. Martinez-Allier, 2002; Bond, 2010; Pickering & Barry, 2012)—should be the ones to foot the bill and take more aggressive climate mitigation action today—voluntary or otherwise. Based on the finance flows supporting Ecuadorian and South African fossil fuel projects, a potential MA may include any of the World Bank and African Development Bank (MDBs), the China Development Bank and Brazil's BNDES (BDBs), and various European governments via their ECA involvement, like Italy's SACE, France's COFACE and Germany's Euler Hermes.

Once established, the MA may adopt any number of effective supply-side measures to ensure that the coal, oil and gas reserves from a particular reserve remain underground. For example, the MA members may implement

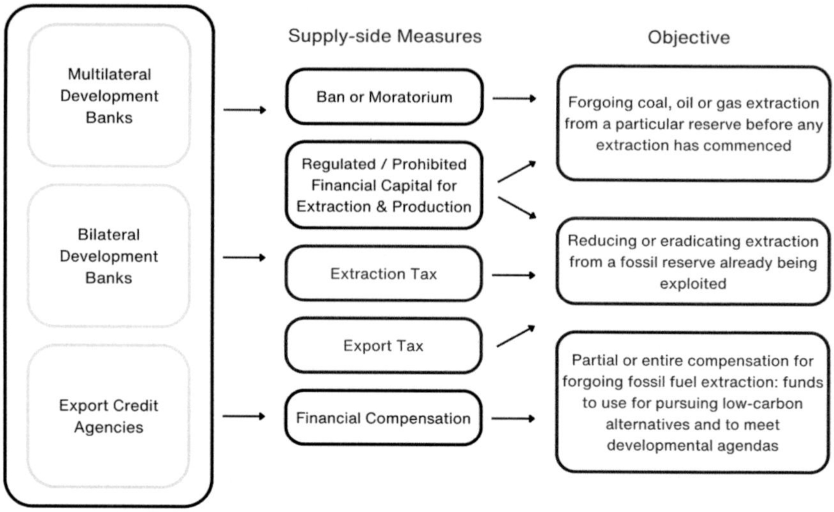

Figure 8.1. Example of a mitigation alliance mechanism.
Source: By the authors.

policies that regulate (or prohibit altogether) available financial capital for extraction or production from the reserve (Best, 2017; Gunningham, 2020). As we discussed in 8.4, without MDB, BDB or ECA support for a project, the risk borne by commercial investors would be far too high and the fossil investment would very likely be forgone.

If extraction is already underway from a reserve the MA can alternatively support the host government in crafting an export or extraction tax (e.g. Sinn, 2012; Lazarus & Van Asselt, 2018; Faehn et al., 2017). Contrary to the traditional (demand-side) emissions tax, these supply-side approaches can disincentivise extraction from a reserve by taxing "the capital income earned on [fossil] assets … to make … natural capital more attractive or, alternatively, make … financial assets less enticing!" (Sinn, 2012, p. 216).

Finally, and perhaps most significantly, the MA can craft a financial package to offer compensation to a government to cover the cost of forgoing fossil fuel development, building on the lessons learned from the shortcomings of the Yasuní–ITT Initiative in Ecuador (Larrea & Warnars, 2009). This financial injection could cover a fraction or even a majority of the estimated present value of the fossil fuel reserves and a legally binding agreement with the recipient nation would be established to ensure that the proceeds are used to pursue their development agenda in alignment with the SDGs. For example, the financial package may fund the development of grid-scale solar PV, wind power and battery storage (SDGs 7 and 9), which would inadvertently

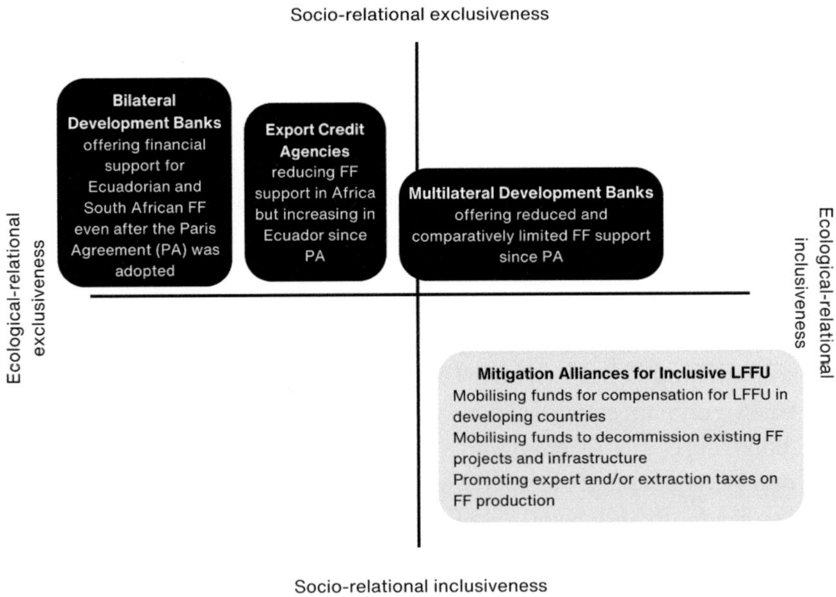

Figure 8.2. Supply-side/institutional mechanisms and inclusive transition.

stimulate domestic employment (SDG 8) or a universal basic income grant to the nation's citizens to alleviate poverty (SDG 1) and hunger (SDG 2), all the while taking effective climate action (SDG 13) and preserving biodiversity hotspots (SDGs 14 and 15). As such, mitigation alliances for inclusive LFFU would combine eco-relational and socio-relational inclusiveness and imply a profound change in the role of PFIs (see Figure 8.2).

A potential MA of the numerous PFIs that have historically financed South Africa's fossil fuel regime (see Table 8.7 and Table 8.8) can take a number of supply-side measures to ensure that these resources remain underground and uncombusted. First and foremost, given that French multinational Total operates the block, the MA—which should include French ECA COFACE—can adopt an immediate policy regulating public finance for developing this reserve in a bid to disincentivise further commercial investment. Furthermore, the MA can jointly allocate funds to compensate (partially or entirely) the South African people for forgoing the opportunity of realising these \$465–558 billion in prospective revenue streams. Given the corrupt history that has and continues to plague the South African government, this financial swap should be administered by the MA itself in collaboration with the South African National Treasury, the Department of Minerals & Energy, the Department of Environment, Forest & Fisheries, and other relevant ministries to ensure the funds are used in accordance with South Africa's development agenda (NPC, 2013). This could include, for instance, allocating funds to decommission old and dilapidated coal-fired power generators in Mpumalanga province and replace them with low-carbon solar PV (SDGs 7 and 9), all the while stimulating local jobs (SDG 8) and providing a basic income to ensure that the province's coal miners and other mining dependents are not left stranded and unemployed (SDGs 1 and 2).

Recently, promising new supply-side initiatives and MAs have been proposed. In South Africa, an agreement to help the country to reduce the use of coal to generate electricity and develop clean energy sources, with an $8.5-billion contribution, mostly from the US, the EU, the UK, Germany, and France, was proposed at COP26 in Glasgow (Pilling, 2022). This may be one of the first significant steps of supply-side mitigation. Meanwhile in Ecuador's neighbouring country Colombia a policy proposal was presented in 2022 by the new government of President Petro. This proposal contains a transition plan to gradually reduce oil extraction and promote a transition to a low-emission economy, including a ban of oil exploration using fracking, and a gradual decline of oil extraction from current fields in the Amazon. This policy may become the first integrated transition policy to move away from fossil fuels in Latin America. Oil is currently the main export in Colombia, and the change involves a significant economic shift (CENSAT, 2022). The energy transition in Colombia may offer an opportunity to apply the policy proposals for LFFU analysed in this chapter and inspire other countries.

8.6. Conclusion

Supply-side approaches for LFFU are prospectively more effective, feasible and cheaper to implement than their demand-side counterparts (e.g. a traditional carbon emissions tax), yet they have been comparatively under-researched (Le Billon & Kristoffersen, 2019; Lazarus & Van Asselt, 2018; Sinn, 2012). Moreover, measures to account for the international finance flows in phasing out fossil fuels have been almost entirely omitted from existing scholarship (Rempel & Gupta, under review; cf. Newell & Simms, 2020), which is in direct misalignment with various pledges in the Paris Agreement, for example, to "mak[e] finance flows consistent with a pathways towards low greenhouse gas emissions and climate-resilient development" (Article 2.1c) and to provide "[s]upport ... to developing country Parties ... allow[ing] for higher ambitions in their actions" (Article 4.5).

Meanwhile, public finance institutions (PFIs)—including multilateral and bilateral development banks (MDBs and BDBs) and export credit agencies

(ECAs)—have profusely financed and continue to finance coal, oil and gas projects across the globe, including in vulnerable Global South regions and countries like Ecuador and South Africa (see 8.4). From 2016 to 2019, finance from G20 PFIs (predominantly from Chinese BDBs, the World Bank and the African Development Bank, and a series of European and East Asian ECAs) to Ecuador and South African fossil fuel projects may have aggregated to over $5 billion. This is in direct violation of the commitments set forth in the Paris Agreement given that the vast majority of coal, oil and gas resources must be left underground and untouched in order to limit average global temperature rise to 1.5 °C (McGlade & Ekins, 2015; Welsby et al., 2021).

To better align with the Paris Agreement and simultaneously take effective supply-side fossil fuel action, we propose crafting mitigation alliances (MAs), composed of governments and financial institutions that have historically enabled fossil projects in developing and vulnerable regions, like the aforementioned PFIs. The MA can subsequently adopt effective supply-side regulatory policies, like banning or prohibiting available finance for developing a particular fossil reserve, or fiscal policies, like compiling a financial package to compensate a nation for forgoing the opportunity of developing or commercialising a fossil reserve—resembling and learning from the shortcomings of the Yasuní–ITT Initiative in Ecuador. This financial injection can be used to promote the nation's development agenda by offering a basic income grant (SDG 1) or stimulating local jobs (SDG 8) by developing low-carbon infrastructure (SDGs 7 and 9), all the while taking effective climate action and protecting critical biodiversity hotspots (SDGs 13–15). Although potentially costly—particularly for the MA—this approach is fully within the bounds of the Paris Agreement and may provide a unique route to an inclusive fossil transition by allocating the stranded asset costs to rich and capable actors within the MA rather than under-resourced and underprivileged citizens from vulnerable regions.

References

Armstrong, C. (2019). Decarbonisation and world poverty: A just transition for fossil fuel exporting countries? *Political Studies, 68*(3), 671–688. https://doi.org/10.1177/0032321719868214

Arsel, M., Hogenboom, B., & Pellegrini, L. (2016). The extractive imperative in Latin America. *The Extractive Industries and Society, 3*(4), 880–887.

Aviva Investors, Investec Asset Management & WWF-UK. (2015). *Safeguarding outstanding natural value: The role of institutional investors in protecting Natural*

World Heritage Sites from extractive activity. https://wwf-sight.org//wp-content/uploads/2017/01/wwf_nwh_investor_report_a4_web_v2_1.pdf

BankTrack. (N.d.). Medupi coal power plant, South Africa. https://www.banktrack.org/project/medupi_coal_power_plant

Bass, M., et al. (2010). Global conservation significance of Ecuador's Yasuní National Park. *PloS ONE, 5*, 1. https://doi.org/10.1371/journal.pone.0008767

Best, R. (2017). Switching towards coal or renewable energy? The effects of financial capital on energy transitions. *Energy Economics, 63*, 75–83. https://doi.org/10.1016/j.eneco.2017.01.019

Bond, P. (2010). Climate debt owed to Africa: What to demand and how to collect? *African Journal of Science, Technology, Innovation and Development, 2*(1).

CENSAT. (2022). Disminución planeada de la dependencia fósil en Colombia. https://co.boell.org/es/2022/10/21/disminucion-planeada-de-la-dependencia-fosil-en-colombia

Coady, D., Parry, N., Le, P., & Shang, B. (2019). *Global fossil fuel subsidies remain large: An update based on country-level estimates.* IMF. https://www.imf.org/en/Publications/WP/Issues/2019/05/02/Global-Fossil-Fuel-Subsidies-Remain-Large-An-Update-Based-on-Country-Level-Estimates-46509

Erickson, P., Lazarus, M., & Piggot, G. (2018). Limiting fossil fuel production as the next big step. *Nature Climate Change, 8*, 1037–1043. https://doi.org/10.1038/s41558-018-0337-0

Faehn, T., Hagem, C., Lindholt, L., Maeland, S., & Rosendahl, K. (2017). Climate policies in a fossil fuel producing country: Demand versus supply side policies. *The Energy Journal, 38*(1), 77–102. https://doi.org/10.5547/01956574.38.1.tfae

Gaulin, N., & Le Billon, P. (2020). Climate change and fossil fuel production cuts: Assessing global supply-side constraints and policy implications. *Climate Policy, 20*(8), 888–901. https://doi.org/10.1080/14693062.2020.1725409

Green, F., & Denniss, R. (2018). Cutting with both arms of the scissors: The economic and political case for restrictive supply-side policies. *Climate Change, 150*, 73–87. https://doi.org/10.1007/s10584-018-2162-x

Gunningham, N. (2020). A quiet revolution: Central banks, financial regulators, and climate finance. *Sustainability, 12*(22), 9596. https://doi.org/10.3390/su12229596

Gupta, J., Rempel, A., & Verrest, H. (2020). Access and allocation: The role of large shareholders and investors in leaving fossil fuels underground. *International Environmental Agreements: Politics, Law and Economics, 20*(2), 303–322. https://doi.org/10.1007/s10784-020-09478-4

IEA [International Energy Agency]. (2012). World energy outlook, executive summary. http://www.worldenergyoutlook.org/publications/

Johnsson, F., Kjärstad, J., & Rootzén, J. (2019). The threat to climate change mitigation posed by the abundance of fossil fuels. *Climate Policy, 19*(2), 258–274. https://doi.org/10.1080/14693062.2018.1483885

Kingsbury, D. V., Kramarz, T., & Jacques, K. (2018). Populism or petrostate? The afterlives of Ecuador's Yasuní–ITT Initiative. *Society & Natural Resources, 32*(5), 530–547. https://doi.org/10.1080/08941920.2018.1530817

Larrea, C., Adoum, A., & Huber, R. (2009). *Yasuní–ITT: An initiative to change history.* Gobierno Nacional de la República del Ecuador.

Larrea, C., et al. (2021). Globalization, extractivism, and social exclusion: Country-specific manifestations. In C. Nobre et al. (Eds.), *Science Panel for the Amazon* (ch. 18). United Nations Sustainable Development Solutions Network. https://doi.org/10.55161/ZMNH8351

Larrea, C., & Warnars, L. (2009). Ecuador's Yasuní–ITT Initiative: Avoiding emissions by keeping petroleum underground. *Energy for Sustainable Development, 13*(3), 219–223. https://doi.org/10.1016/j.esd.2009.08.003

Lazarus, M., Erickson, P., & Tempest, P. (2015). *Supply-side climate policy: The road less taken.* Stockholm Environment Institute. https://mediamanager.sei.org/documents/Publications/Climate/SEI-WP-2015-13-Supply-side-climate-policy.pdf

Lazarus, M., & Van Asselt, H. (2018). Fossil fuel supply and climate policy: Exploring the road less taken. *Climate Change, 150*, 1–13. https://doi.org/10.1007/s10584-018-2266-3

Le Billon, P., & Kristoffersen, B. (2019). Just cuts for fossil fuels? Supply-side carbon constraints and energy transition. *Environment and Planning A: Economy and Space, 52*(6), 1072–1092. https://doi.org/10.1177/0308518X18816702

Martin, P. L., & Scholz, I. (2014). Policy debate: Ecuador's Yasuní–ITT Initiative: What can we learn from its failure? *International Development Policy/Revue internationale de politique de développement, 5*(2). https://doi.org/10.4000/poldev.1705

Martinez-Alier, J. (2002). The ecological debt. *Kurswechsel, 4*, 5–17. http://www.beigewum.at/wordpress/wp-content/uploads/005_joan_martinez-alier.pdf

McGlade, C., & Ekins, P. (2015). The geographical distribution of fossil fuels unused when limiting global warming to 2 °C. *Nature, 517*, 187–190. https://doi.org/10.1038/nature14016

Meinshausen, M., et al. (2009). Greenhouse-gas emission targets for limiting global warming to 2 °C, *Nature, 458*(7242), 1158–1162. https://doi.org/10.1038/nature08017

Munda, G. (2008). *Social multi-criteria evaluation for a sustainable economy.* Springer.

Newell, P., & Simms, A. (2020). Towards a fossil fuel non-proliferation treaty. *Climate Policy, 20*(8), 1043–1054. https://doi.org/10.1080/14693062.2019.1636759

NPC [National Planning Commission]. (2013). *National development plan vision 2030.* Republic of South Africa.

OCI [Oil Change International]. (2020). Shift the Subsidies Database. http://priceofoil.org/shift-the-subsidies/

Pellegrini, L., Arsel, M., Orta-Martínez, M., Mena, C. F., & Muñoa, G. (2021). Institutional mechanisms to keep unburnable fossil fuel reserves in the soil. *Energy Policy, 149*, 112029.

Pickering, J., & Barry, C. (2012). On the concept of climate debt: Its moral and political value. *Critical Review of International Social and Political Philosophy, 15*(5), 667–685. https://doi.org/10.1080/13698230.2012.727311

Pilling, D. (2022, November 22). The cost of getting South Africa to stop using coal. *Financial Times*.

Rempel, A. (2023). An unsettled "stranded asset debt"? Proposing a supply-side counterpart to the "climate debt" in a bid to guide a just transition from fossil fuels in South Africa and beyond. *Antipode, 55*(1), 243–267.

Rempel, A., & Gupta, J. (2022). Equitable, effective, and feasible approaches for a prospective fossil fuel transition. *WIREs Climate Change, 13*(2), e756. https://doi.org/10.1002/wcc.756

Rempel, A., & Gupta, J. (Under review). Public finance and (African) fossil fuels: Export credit agencies, development banks and the Paris Agreement. *Global Environmental Change*.

Sinn, H. W. (2012). *The green paradox: A supply-side approach to global warming.* MIT Press.

Spalding, R. J. (2018). From the streets to the chamber: Social movements and the mining ban in El Salvador. *European Review of Latin American and Caribbean Studies/Revista europea de estudios latinoamericanos y del Caribe, 106,* 47–74.

Steffen, W., et al. (2015). Planetary boundaries: Guiding human development on a changing planet. *Science 347*(6223), 736. https://doi.org/10.1126/science.1259855

Stua, M. (2017). *From the Paris Agreement to a low-carbon Bretton Woods.* Springer.

Tudela, F. (2020). Obstacles and opportunities for moratoria on oil and gas exploration or extraction in Latin America and the Caribbean. *Climate Policy, 20*(8), 922–930.

UNEP [United Nations Environment Programme]. (2022). *The closing window: Climate crisis calls for rapid transformation of societies.* https://www.unep.org/resources/emissions-gap-report-2022

UNFCCC [United Nations Framework Convention on Climate Change]. (2015). Adoption of the Paris Agreement. Report No. FCCC/CP/2015/L.9/Rev.1. Conference of the Parties 21st Session. Paris. https://unfccc.int/sites/default/files/english_paris_agreement.pdf.

Vallejo, C., Burbano, R., Falconí, F., & Larrea, C. (2015). Leaving oil underground in Ecuador: The Yasuní–ITT Initiative from a multi-criteria perspective. *Ecological Economics, 109,* 175–185. http://doi.org/10.1016/j.ecolecon.2014.11.013

Welsby, D., Price, J., Pye, S., & Ekins, P. (2021). Unextractable fossil fuels in a 1.5 °C world. *Nature, 597,* 230–234. https://doi.org/10.1038/s41586-021-03821-8

World Bank. (2022). *State and trends in carbon pricing 2022.* https://openknowledge.worldbank.org/handle/10986/37455

9. Phasing out Fossil Fuels: Synergies and Trade-offs

Joyeeta Gupta, Arthur Rempel and Malin Olofsson

Abstract

This chapter concludes by answering the questions: How do different actors use different arguments and approaches to engage with a fossil fuel phase-out? Which arguments and approaches grapple with the multiple trade-offs, for which actors, and why? And how can these be scaled up? The chapter identifies 10 niche arguments that need scaling up: Revisiting how development is defined, using court cases as an effective option for social movements, social movements can build on existing contextual ideas such as the rights of nature, the need for extraction moratoria, demanding compensation for stranded fossil fuel assets, emphasizing the vulnerability of exports to border tax adjustments can influence national and business policy, holding all accountable for their own climate debt, subject fiduciary responsibility to socio-ecological criteria, use philanthropic funds to decommission existing fossil fuel projects and adapt and implement OECD policy to prohibit ECA finance for fossil fuel projects.

Keywords: fossil fuel phase-out, scaling-up, arguments, approaches, inclusive development

9.1. Introduction

We have looked at the challenges of leaving fossil fuels underground through examining the different arguments and approaches that are used by different actors at different levels of governance. We focused on three levels—global

Gupta, J., Hogenboom, B., Rempel, A. & Olofsson, M. (eds), *Leaving Fossil Fuels Underground: Actors, Arguments and Approaches in the Global South and Global North.*
Taylor & Francis Group, 2024
doi: 10.5117/9789048560370_CH09

actors, regional actors and local actors in South Africa and Ecuador—posing the tri-pronged question: *How do different actors use different arguments and approaches to engage with a fossil fuel phase-out? Which arguments and approaches grapple with the multiple trade-offs, for which actors, and why? And how can these be scaled up?* In responding to this question, we have focused less on the fact that there is a huge global fossil fuel lock-in where powerful actors and national governments are highly motivated to delay the fossil fuel phase-out. Instead we have focused on positive narratives of social movements and other parties that can mobilise a global challenge to the continued use of fossil fuels.

This chapter unites the insights that have emerged from the different chapters, by linking them with the Sustainable Development Goals (see Chapter 1) and by identifying which niche activities can be scaled up to the regime and landscape levels (see Chapter 2). Our different chapters looked at various issues from diverging perspectives—from the role of social movements and courts more broadly in the Global South (see Chapter 3), to social movements and activities initiated in Latin America including Ecuador (see Chapter 4), the policy challenges in South Africa (see Chapter 5); social mobilisation and coalition forming in South Africa (see Chapter 6); the role of finance for fossil fuel investments in the Global South in the fossil fuel narrative (see Chapter 7) and the role of supply-side initiatives (see Chapter 8).

We now revisit the challenge of climate change, energy and the posed implications for development and justice. In doing so, we first elaborate on the diversity of actors, arguments and approaches that are involved in fossil fuels (see 9.1). This brings us to a discussion of environmental justice (fighting against injustice) versus legal justice (weighing the arguments of all actors using a justice perspective) issues. This requires us to move towards the inclusive development frame which uses the lens of social, ecological and relational issues to examine and critique economic issues (see 9.2). We then link this to the Sustainable Development Goals building on our framework in Chapter 1 to highlight the synergies and conflicts (see 9.3). Such a frame also enables us to identify key ideas, narratives and tools that need to be scaled up to enable a just process towards leaving fossil fuels underground (see 9.4). Finally, we integrate our disparate strands of analysis into our analysis of transition processes (see 9.5) before drawing conclusions.

We acknowledge that we take a broad North–South approach in this book, while we recognise that there are many nuances to the storyline, which we bring out as and when we see that as necessary.

9.2. Actors, arguments and approaches

Our research has focused on three key actors—social movements, the large investors and governments in the Global South. We did not look at other actors such as labour movements and epistemic communities but complement our analysis in this section by building on the relevant literature.

9.2.1. Social movements and NGOs

A key actor at local to global level are NGOs and social movements. NGOs are organised formal bodies with a secretariat while social movements coalesce around specific themes. The latter can be generated by NGOs or they may grow to include NGOs. Their arguments with respect to fossil fuel can be clustered into four storylines: (a) the impacts of extraction, production and use affects local communities either through a direct loss of livelihoods or through damaging their environment and resources and thereby also affecting their health; this has been used to mobilise grievances locally to create a social movement (see Chapters 3 and 6); however, this can also be affected by loss of jobs in the fossil fuel sector; (b) these primary arguments have been supported by other arguments including that of the global problem of climate change; climate change has not been the motivating arguments in these movements; (c) these arguments have been used also to merge with other ongoing struggles of peasants, Indigenous communities and women (see Chapter 4); and (d) some social movements have revolved around specific ideas such as the Yasuní–ITT Initiative (see Chapters 1, 3, 8) and the climate debt of countries (see Chapter 5); where the former focuses on receiving compensation from the North for leaving fossil fuels underground and the latter on the climate debt incurred by taking fossil fuels out.

In terms of approaches, social movements have engaged in activities ranging from public education campaigns, naming and shaming, lobbying and litigation, and violence/"blockadia" activities in relation to LFFU. Education campaigns focus on mobilising critique of dominant development paradigms, neo-extractivist strategies, economic and ecological debt and the promotion of alternative visions. Lobbying and litigation to mobilise state and other actors have included demanding the closure of fossil fuel activities and change in fossil fuel policy. The former takes precedence over the latter. Such activities have taken place where there is civic space in more democratic countries. Where there is less civic space, petro-violence and "blockadia" activities have been more dominant but there is less evidence as to whether this actually leads to a change in policy by itself (see Chapters 3–6).

Some social movements in the Global South have been successful, such as the public demand for a moratorium on oil exploration and extraction as in Costa Rica and the petroleum moratorium in Belize waters. Social movements tend to be successful when they can mobilise local people by focusing on their local and immediate interests such as their livelihoods, their environment and their health. They can also make links to existing grievances of these communities. Global issues then piggyback on such interests. However, social movements tend to be more successful when the economic losses of local people outweigh the economic losses of the other party—a point we return to later.

Where social movements and other actors have used the courts, they have often been more successful in closing down fossil fuel projects and/or in demanding policy change (see Chapter 4). Factors influencing the use of courts in the Global South include the degree to which a society is democratic (which enables actors to question state action), the degree to which a society has developed institutions (which ensures a functioning judiciary and the ability of social actors to prepare a court case), and the degree to which a country emits greenhouse gases (which creates a stronger need for action). We did not find a clear link between countries that were more or less vulnerable to the impacts of climate change and the number of court cases.

9.2.2. Investors and financiers

We now turn to another key actor—those who both finance and invest in fossil fuel projects. The former typically concerns financial bodies that issue credit (i.e. loans) to companies or projects that drive fossil fuel extraction or production projects; these can include commercial banks, development banks or various other types of public finance institutions (PFIs). Meanwhile, the latter typically refers to shareholders that make equity investments in companies that produce fossil fuels; these can include pension funds, insurance companies or several other types of institutional investors. In this book, we have studied philanthropic organisations, pension funds, and investor initiatives in terms of investors, and export credit agencies (ECAs).

We find that while philanthropic organisations admittedly aim at spending their grants on some climate mitigation and adaptation projects that ostensibly adhere to the Paris Agreement objectives, the source of this grant funding (i.e. their investment portfolios) contradictorily does not exclude or prohibit investments in fossil fuel extraction and production firms, revealing a concerning incoherence between their investment and philanthropic agendas. This therefore questions the extent to which philanthropic foundations are to any degree effective in promoting a fossil fuel phase-out; that is, how can an organisation successfully support a cause, such as climate mitigation via LFFU, if the means by which it supports this cause relies (potentially quite heavily) on investments that paradoxically drive and exacerbate the very problems they aim to "solve"? Our research has made but a small dent in studying how philanthropic foundation investments resonate with or destructively inhibit their missions and purposes; further research is urgently merited, perhaps using more in-depth case study methodology to unpack the nuances and justifications for these apparent discrepancies. This is particularly vital given how the "philanthrocapitalism" paradigm continues to make headlines around the world (Giridharadas, 2018); with figures like Jeff Bezos "donating $10 billion to a climate fund" (Clifford, 2021) should we be rejoicing, or is this all a façade?

Regarding pension funds, their primary and self-assigned responsibility is their fiduciary responsibility to their shareholders; that is, their investment decisions are first and foremost driven and justified by the extent to which such decisions will benefit or detract from the *financial* well-being of their existing and future pensioners. Although many pension funds have established sustainability offices and have pledged to align their investment activities with some form of environmental, social and governance agendas, their outdated internalisation of this fiduciary duty has rendered them incapable of taking effective climate mitigation action that could result in a loss (or simply not as high of a gain) for their shareholders. Debates about whether pension funds (and other institutional investors) should divest from or engage with fossil fuel companies have been running rampant for at least a decade, but our research shows that not only are these approaches ineffective for LFFU (Rempel & Gupta, 2021), but that these actors are not even willing to entertain them to greater extents if it means sacrificing a return on their investment. This poses a dilemma, particularly given that effective climate mitigation via LFFU cannot be fallaciously understood as a "win-win"; there is a bill to pay to phase out fossil fuel dependence, and someone(s) must "lose" in paying this bill—who will it be?

Investor initiatives were subsequently explored, which differ slightly from the above two actor groups, mainly because they are themselves coalitions of investors that may include both pension funds and philanthropic organisations. However, we could see these initiatives as stand-alone actors, namely because of the power that they may wield in swaying fossil fuel firm behaviour by conglomerating their assets under management and exploiting their joint leverage over the companies in which they invest. Although this is a tremendously under-explored field that merits much more forthcoming research, our preliminary study suggests that these initiatives are not even beginning to make the best use of their potential, and have thus far played more of a greenwashing role than anything else, forming the illusion that institutional investors are taking effective climate action, while in reality they allow said investors to delay phasing out fossil fuels as they scramble to decide to whom to pass on the hot potato of fossil fuel assets that may suddenly evaporate.

Export credit agencies (ECAs) are a unique type of PFI; they are typically one financial arm of a government, and financially support domestic companies who seek to conduct business abroad by offering either credit (i.e. loans, debt) or guarantees (i.e. insurance). Their role in doing so cannot be sufficiently stressed; if commercial banks are uncomfortable with the financial risks posed by a new fossil fuel project, an ECA (or other PFI, for that matter) can step in and de-risk the project, mobilising finance for a project that may not have otherwise been available. Most OECD ECAs have continued to fund fossil fuel projects abroad, particularly in the Global South, though this is on the decline—particularly in the case of thermal coal mining and combustion projects. This effectively means that between 2010 and 2020, ECAs from the Global North have hoisted and generated fossil fuel infrastructure in the Global South—the very infrastructure that is incompatible with the objectives of the Paris Agreement and must be decommissioned as soon as possible to avoid further lock-in in the Global South. There is considerable work to be done in ensuring financial coherence in this field as this is not a simple task. Not making new investments in fossil fuels is the easiest. But those who have financial stakes in the fossil fuel sector have two choices—to phase out such investments by selling the stakes or to engage with their debtors to ensure that they transition out of fossil fuels. Each of these choices is problematic—the first ensures that the financer is stable, but creates new vested interests; and the second shifts the responsibility to the party borrowing the money or invested in.

9.2.3. Governments

The government narrative on the side of the Global South can be clustered into three storylines. The first, is the multidimensional sacrifices (e.g. the loss of energy, jobs, income, and spillover effects on the economy) involved in leaving fossil fuels underground which either imply investing in fossil fuels and making profits in the short to medium term (as is illustrated in Chapter 5); not investing in the fossil fuel sector and calling for compensation as is demonstrated in Chapter 8; and finally not investing in fossil fuels and accepting that it will be a stranded asset as many social movements appear to want (see Chapters 3, 4 and 6). All this falls under the right to development. The second, is the strategic adoption of key demand and supply-side policies to LFFU—which implies focusing on improving energy efficiency and shifting to alternative fuels; and, third is the fear that trade restrictions will force a shift away from fossil fuels as is illustrated in Chapter 5. Some of the choices in South Africa reveal corrupt practices of vested interests in government where those who have interests in the fossil fuel sector try to mobilise policy in their favour (see Chapter 5).

Foreign governments from the European Union, Japan and the United States are potential key actors—by being financers of projects (e.g. through export credit) and by using border tax adjustments. Such border tax adjustments could affect South Africa's (and all other developing countries') abilities to engage in trade with the European Union or other country that applies such an adjustment when the products and services exported are produced using fossil fuels. The government narrative on the side of the Global North can be clustered around: (a) limiting the risk of stranded LFFU assets through e.g. continued use of fossil fuels at home and export credit to fossil fuel firms; (b) the strategic use of demand and supply-side options, and (c) using trade restrictions to force the rest of the world into compliance. With the war on Ukraine (see Box 9.1) and reduced supplies of fossil energy to the Global North, there appears alas to be a return to investments in, and subsidies for, fossil fuels. The choices in the Global North may not necessarily be attributed to corruption but do reveal incoherent policies. This raises questions regarding whether the policy results from a necessity to invest in fossil fuels, or whether the vested interests in the fossil fuel sector are shaping national policy—either directly by being in political positions or indirectly through the use of their power and possibly pay-offs. Those with vested interests in the fossil fuel sector are also powerful and will use legal and quasi-legal ways to ensure the future of fossil fuels.

Box 9.1. The war against Ukraine

Russia's war against Ukraine since February 2022 has caused a global energy crisis, with demand exceeding supply and a sharp increase in prices worldwide (Tollefson, 2022). As the largest exporter of natural gas and second-largest exporter of crude oil globally (UN Global Crisis Response Group on Food, Energy and Finance, 2022), the European Union depended on Russia for approximately 40% of its oil and gas supply (Campbell, 2022a and 2022b), and other European countries were more heavily reliant (Tollefson, 2022). However, as European countries imposed economic sanctions on Russia (Obisie-Orlu, 2022) this necessitated subsidising energy prices and searching for alternative energy sources to close the gap (Butler, 2022), for example, through oil and gas imports from Qatar, Algeria (Butler, 2022), Israel (Ghiles, 2022), Egypt (Fisher, 2022), Mozambique, Libya, and Morocco (Ghiles, 2022), Angola (Campbell, 2022a), Senegal (Vyshnytska, 2022), Congo (Campbell, 2022a), Azerbaijan (Kuzemko et al., 2022), Nigeria (Ruta, n.d.) and Mexico (Chepeliev et al., 2022); and thermal coal imports from South Africa and Indonesia, which increased more than 11-fold since the war (Kumar & Levitan, 2022).

Consequently, the war may delay the clean energy transition in the Global South (Tollefson, 2022), "not least by creating incentives, directly or indirectly, for investment in coal, gas and associated supply chains" (Kuzemko et al., 2022). In this regard, a recent Memorandum of Understanding between Afreximbank and the African Petroleum Producers' Organization (APPO) aims to create an African Energy Transition Bank to finance oil and gas projects on the continent (Obisie-Orlu, 2022), while other actors may invest as much as $100 billion in oil and gas projects in Africa (Kumar & Levitan, 2022). The war may, thus, ironically send signals for increasing, rather than decreasing, investments in fossil fuels. Further, Russia is now exporting its excess supply of fossil fuels to the Global South such as India, China and Turkey, while at least during the initial months, Japan, South Korea, and Taiwan maintained significant imports of fossil fuels from Russia (Nadig, 2022).

9.2.4. Implications for environmental justice, legal justice and inclusive development

Thus, there are a range of narratives that are brought forward by different actors. Here we make the following points: (a) we differentiate between environmental justice and legal justice; (b) we argue that legal justice may weigh in favour of economic arguments; (c) unless we are able to convincingly use an inclusive development and justice framework to re-examine the economic arguments.

Environmental justice movements protest against injustices; they fight against the way in which environmental damage affects their current or future lives and livelihoods. Using the courts is an effective way for them to ensure a *niche* change, and maybe even a *regime* change. However, courts look at justice in a slightly different manner; they weigh the pros and cons of the justice issues related to all actors. They will look at the arguments of social movements, scientists, labour groups, investors and so on to come to a balanced decision based on all the issues concerned. In such a decision, a key question is: Will the economic losses prevail over the socio-ecological gains of LFFU; or will science and social movements prevail against the economic issues? The courts will weigh the trade-offs mentioned above.

And this brings us to a discussion of justice and inclusive development. If social, ecological and economic issues are to be weighed against each other, the short- to medium-term economic issues will prevail. Inclusive development therefore calls for weighing social, ecological and relational issues against each other and then looking at the economic dimensions (see Chapter 2); in other words, inclusiveness is not an adjective to development, it redefines development (Rammelt & Gupta, 2021).

Inclusive development calls for examining social, ecological and relational issues. In relation to each we focus on just access and just allocation. Within the *social* domain, just access implies looking at (i) income/jobs above the poverty line; (ii) access to water, food, energy, health, housing and infrastructure; (iii) access to procedural rights including non-discrimination, rights of minorities, women, developing countries; and rights to information, decision-making, recognition and courts. Just allocation examines (i) socio-economic resources (e.g. fossil fuel resources) and infrastructures; (ii) socio-economic risks; and (iii) socio-economic responsibilities.

Within the *ecological* domain, just access includes NCPs—material, non-material, regulating (all with a cultural component)—for the most marginalised to be able to access (i) income; (ii) healthy lives; and (iii) environmental rights. Just allocation refers to the distribution of the: (i) remaining NCPs; (ii) related risks; and (iii) environmental responsibilities.

Within the *relational* domain, we are examining the role of states (and interstate relations) and their ability to provide and/or protect social and ecological common goods, and to address inequalities; the role of other actors, including financial actors and their desire for maintaining the status quo, and social movements and their desire for change; and the role of dominant and alternative approaches and arguments in shaping the adoption of substantive and procedural principles, rights and responsibilities to address socio-ecological justice. These elements have been combined for our analysis below in Section 9.3. Since such inclusiveness requires redefining development, we return to this in Section 9.4.

9.3. Multidimensional trade-offs

In Chapter 1, we identified four clusters of issues that could be relevant for looking at the implications of LFFU and the SDGs. We now discuss these in the light of some of the arguments that have surfaced throughout this research. This is based on looking at justice in terms of *access* to basic needs, as specified in the SDGs, and the *allocation* of resources, risks and responsibilities (Gupta & Lebel, 2020a, 2020b).

9.3.1. Basic resources and jobs

First, in terms of jobs/income, social movements have mobilised grievances and tended to protest against the impacts of energy-related pollution on the environment which then affected their livelihood prospects (see Chapters 3 and 6). Since in many countries there has been no active discussion of phasing out existing fossil fuel plants, but rather the emphasis has been on new extraction and new to-be-built fossil fuel plants, workers in the fossil fuel industry have not yet protested actively. However, if such a phase-out begins, there will be a major backlash from the labour movements and calls for a just labour transition are growing worldwide (see Chapter 6).

9.3.2. Energy

Second, the SDGs aim to ensure that people, especially those living in energy poverty, have access to affordable, reliable and modern energy. This narrative did not emerge strongly in our research, but this may have been because (a) we did not conduct interviews with those with limited access; (b) these people are not organised within social movements demanding energy access; and (c) there is an assumption that renewable energy will provide the decentralised access to affordable, reliable and modern energy for these communities. (Our field work has been seriously affected by COVID-19.) While the latter is certainly the hope, there is considerable doubt about the availability of battery systems to ensure access to renewables at affordable prices for people worldwide. In our mind, there is no doubt that this is a key social issue that needs to be addressed.

9.3.3. Nature's contributions to people

Third, when discussing nature's contributions to people (NCP), social movements are organising to protest against the damage caused by fossil fuel-related activities to their environment. However, it is clear that investors, producers, users and governments see that the economic benefits of using the provisioning services of these resources outweighs the damage to the supporting, regulating and cultural services that these resources provide.

9.3.4. Procedural issues

Fourth, in terms of procedural issues, local actors and social movements increasingly have access to information and courts (at least in democratic countries), if not to processes of decision-making. These have often been complemented by petro-violence and "blockadia" approaches. The fossil fuel labour movements appear not to have organised themselves yet. Energy-poor communities also are not yet organised to demand procedural rights.

We then look at issues related to socio-ecological *allocation* in terms of inequality; production and consumption patterns; North–South issues; and the right to development.

9.3.5. Inequality

Energy use has been critical to the development of Western society since the Industrial Revolution and the development of the steam engine. While historically these gains were not equitably shared in society as illustrated by the books of Charles Dickens among others, with growing democracy, the gains from energy use have been shared in many parts of the industrialised world—for example in the Netherlands and Norway. However, in the developing world the paradox of plenty has often been a resource curse causing uneven development and exacerbating inequality. It has often led to state monopolies or collusion with companies in corporate endeavours that could often be labelled as state corporate "corruption." Furthermore, more research needs to be undertaken on the limits to energy use, the overuse of energy resources by some and the underuse by others. However, our research shows that shareholders in the energy sector in Africa are predominantly from other parts of the world (except in the case of South Africa; see Chapter 5). This would imply that the profits from using fossil fuel flow elsewhere while the risks are borne by local communities.

9.3.6. Production and consumption

Our research in the Global South shows that production, consumption and the use of fossil fuel resources is on the rise, as is their use of renewables. This shows a growing fossil fuel lock-in in the Global South; moreover, loans for financing such extraction are being taken from international banks and this could also imply a growing debt crisis. The war on Ukraine has led to increased investments in the Global South in fossil fuels. Since companies have instrumental, structural and discursive power (see Chapter 2), they are in a better position to control the agenda.

9.3.7. North–South issues

We took a limited look at the role of Western countries in terms of their investments and production and consumption patterns vis-à-vis the South and conclude that Western investors (a) continue to invest in fossil fuels in the Global South, although this is decreasing; (b) continue to dump know-how, technology and infrastructure onto the South now that it is clear that fossil fuels need to be phased out (see Chapter 7); (c) continue to extract resources from the Global South (the neocolonial disease à la Coronil [2008]); and (d) extract profits from their fossil fuel investments in the Global South. This is very much in line with dependency theory, which argues that the South is both a source of resources and a dumping ground of old technologies and approaches. Some countries in the Global South are also behaving in a similarly predatory fashion—but this is an area of further research.

9.3.8. The right to development

The right to development and energy did not emerge as a dominant theme in this research; however, there is a growing literature on this theme and our regular meetings with African stakeholders tended to emphasise this. Essentially, energy is needed for development; in the past countries used their fossil fuels to become rich quickly. Given that the bulk of the remaining fossil fuels is in the developing world and up to 80% of the fossil fuels need to be kept underground, this de facto leaves the burden of leaving fossil fuels underground to the developing world; this leaves them with massive stranded resources and assets. The global community had begun to discuss how greenhouse gas emissions should be shared between countries but since the Conference of the Parties in Copenhagen in 2009, this discussion has been shelved in favour of allowing countries to voluntarily determine what they will do. In the meanwhile, the window of opportunity for reducing emissions is shrinking fast, de facto, shifting the responsibility for such action to the developing world. This has been accompanied by the "right to promote sustainable development" in the Climate Change Convention (UNFCCC, 1992). It is no wonder then that the developing countries have retaliated by demanding the "right to development" in the Paris Agreement (2015) and Agenda 2030 (UNGA, 2015) as also "full permanent sovereignty" which in theory guarantees them the right to use their fossil fuels under resource nationalism arguments. However, in our view, there needs to be further discussion between countries to achieve a more just result, because there is no way to ensure that developing countries close their fossil fuel enterprises.

In terms of economic issues, this research shows that the short- to medium-term economic value of the fossil fuel enterprise is far in excess of the perceived damage to nations as a whole of shifting to a non-fossil-fuel world. This requires a different perspective on how to govern such a phase-out (see Table 9.1).

We conclude that the most significant synergies in phasing out fossil fuels arises from the social and environmental movements that build on the environmental damage of fossil fuels to their local context and their livelihoods. However, there are far greater trade-offs involved. These trade-offs (a) involve the need for quick, non-intermittent, reliable and affordable energy that can power an economy and raise huge income for a state and that underlies resource nationalism in all countries whether from the Global South or North. Sometimes (b) these nations have taken out huge loans that have to be paid back and that makes it difficult to phase out these activities. Sometimes (c) these nations are influenced by rich actors who are from the fossil fuel industry, and this can include both pressure that is both legal and activities that are extra-legal or corrupt. At the same time, (d) there is a huge army of employees in the fossil fuel sector who run the risk of unemployment if the sector collapses. Finding alternatives for these people in the areas in which they live may not be easy. And communities, households, commuters and others are dependent daily on the fossil fuel economy to cook their food, power their lamps and bring their produce to the market. Revamping the system will not be easy. Our social movement success stories show that it is easier to mobilise policy change at national level when the economic interests of the social movement is higher than the economic interest of the producing partners at the time of discussion (see Chapter 3; Muttitt & Kartha, 2020, p. 4).

Table 9.1. Actors, arguments and approaches and the Sustainable Development Goals

		Issues	**Actors, arguments, approaches**
Justice issues	Socio-ecological access	Jobs/ income (energy related)	Social movements protest against the impact of fossil fuels on their livelihoods and are successful when these economic losses outweigh other economic concerns; Labour movements are likely to resist losing jobs in the fossil fuel sector, there is growing policy concern on this issue, but more research needs to be undertaken on a just labour transition.
		Services (affordable, re- liable, modern energy)	People in energy poverty need such access; they are not organised in social movements, nor do they go to courts for the violation of their energy access rights (as they do for water and sanitation rights); the unanswered question is whether there is enough potential in renewable energy to meet the needs of the poor in a decentralised manner.
		NCPs (energy extraction, production, use & impacts on NCPs)	Social movements protest against fossil fuel extraction, production and use because of the negative impacts on their resources and consequently their health. Investors, producers, users, and governments apparently weigh the use of provisioning resources higher than the impacts on their regulating, supporting and cultural services.
		Procedural rights	Communities and social movements have access to information and courts (at least in democratic societies), if not to decision-making processes. Labour movements appear not to have organised themselves yet. Energy-poor communities also seem to not have availed of their rights to energy.
	Socio-ecological allocation	Inequality (energy related inequalities)	The gains from energy use have been shared in many parts of the industrialised world; but in the developing world the paradox of plenty has often been a resource curse causing uneven development and exacerbating inequality. Furthermore, more research needs to be undertaken on the limits to energy use, the overuse of energy resources by some and the underuse by others. However, our research shows that shareholders in the energy sector in Africa are predominantly from other parts of the world (except in the case of South Africa). This would imply that the profits from using fossil fuels flow elsewhere while the risks are borne by local communities.
		Production/ consumption/ use patterns	Our research in the Global South shows that production, consumption and use of fossil fuel resources is on the rise, as is their use of renewables. This shows a growing lock-in in the Global South; moreover, loans for financing such extraction are being taken from international banks and this could also imply a growing debt crisis.
		North–South energy issues	We examined Western countries only in relation to the investments, production and consumption patterns in the South and conclude that Western investors continue to invest in fossil fuels in the Global South.
		Right to energy & development	The right to development and energy did not emerge as a dominant theme in this research; however, there is a growing literature on this theme.
	Economy	Economy & energy	This research shows that the short- to medium-term economic value of the fossil fuel enterprise is far in excess of the perceived damage to nations as a whole of shifting to a non-fossil-fuel world. This is seen as resource nationalism. This requires a different perspective on how to govern such a phase-out.

9.4. Scaling-up approaches and arguments

This book has raised a number of ideas and approaches that could be scaled up (or rather, pursued) to promote more effective climate action via LFFU. We discuss these below.

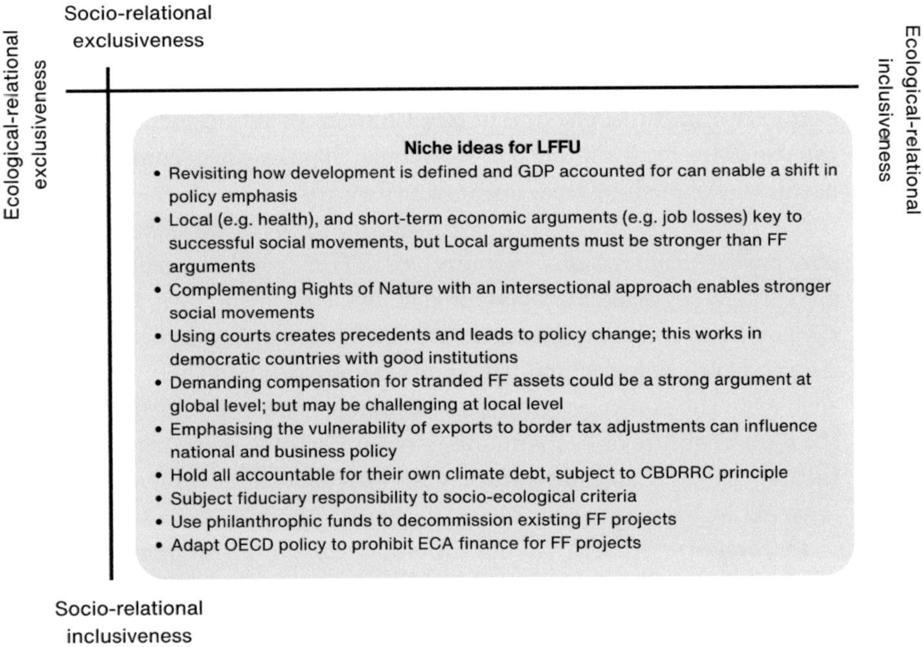

Figure 9.1. Socially and ecologically inclusive niche ideas.

Revisiting how development is defined and GDP accounted for can enable a shift in policy emphasis. There is growing evidence that infinite growth is not possible, and that GDP is poor in representing social and ecological damage. Despite this, countries in the Global North and South emphasise GDP growth and as long as that is the single most important indicator of a country, countries may continue to invest in fossil fuels. There are many alternatives to GDP; culture-based alternatives such as *ubuntu*, *buen vivir* and gross national happiness, but also more science-based alternatives such as inclusive wealth and inclusive development. The concept of *buen vivir* has been adopted in the Constitution of Ecuador and promotes harmonious coexistence with nature. What is critical is that these are not colonised by dominant neoliberal capitalist thinking.

Local (e.g. health) and short-term economic arguments (e.g. job losses key to successful social movements), but local arguments must be stronger than fossil fuel arguments (mobilising grievances). Given the dominance of neoliberal capitalist thinking, it is vital that social and environmental leaders and movements learn from each other. Chapter 3 presents a range of lessons can enable such leaders to mobilise grievances. Such movements need to build on other existing social and environmental movements so that they are not single issue movements but have a more comprehensive story. In Latin America the movements are coalescing around issues of gender, ethnicity and the environment (see Chapter 4). While such movements can use extra-legal approaches, we believe that the use of courts is a more useful way of pushing their arguments forward.

Complementing the rights of nature with an intersectional approach enables stronger social movements. In many parts of the world, the rights of nature are being recognised by social movements, courts and even the law. The right of the river has been recognised in different contexts in Latin America, Asia and even New Zealand. The rights of nature are also being recognised such as in the 2008 Constitution of Ecuador where Article 71 recognises the right of nature or Pachamama (Mother Earth) to exist, regenerate and fulfil its cycles. Combining these with other social movements creates a large public base.

Extraction moratoria. Across the world, there is a rise in moratoria on extraction of different kinds of fossil fuels. In the developing world, we see such decisions banning various kinds of fossil fuel extraction (as in Argentinian municipalities, Costa Rica and Belize) (see Chapters 3 and 4).

State–corporate corruption and crime. This book has hinted at state–corporate corruption and crime. In some countries such collusion is seen as explicitly corrupt, in other countries such collusion is often condoned. We do not see state–corporate collusion as something exclusive to the developing world. There is considerable collusion between large companies and states in the industrialised countries and these activities need to be put under the research spotlight.

Using courts creates precedents and leads to policy change; this works in democratic countries with good institutions. Demanding compensation for stranded fossil fuel assets could be a strong argument at global level but may be challenging at local level. For many developing countries, the right to development combined with the fact that they have scarcely emitted greenhouse gases so far leads them to call for climate compensation. Similar arguments have been made in the past by academics, the OPEC countries and others. The argument of the industrialised countries is that there is no end to making such compensation in terms of both the number of countries and in terms of the number of countries to which such compensation may be due. This demand has been made at COP27; the question is: Will there be enough funds to compensate for all the loss and damage? Some of our chapters have shown that creative thinking about the role of the large investors could find new funds for such compensation.

Emphasising the vulnerability of exports to border tax adjustments can influence national and business policy. Countries that export heavily to the EU may lose income if border tax adjustments become a reality. This could be used as an argument to convince such countries to move away from fossil fuels.

Hold all accountable for their own climate debt, subject to the common but differentiated responsibilities (CBDR) principle. Every country that continues to emit greenhouse gases creates a climate debt. Chapter 5 shows what the climate debt of South Africa looks like. The notion that every country has some degree of liability vis-à-vis others may change the cost-benefit analysis that many countries undertake in their analysis of energy.

Subject fiduciary responsibility to socio-ecological criteria. Many investors claim that fiduciary responsibility to their shareholders precludes active consideration of the socio-ecological criteria. The question is whether social movements should demand that such fiduciary responsibilities should be made subject to such criteria.

Use philanthropic funds to decommission existing fossil fuel projects. Philanthropies often earn from fossil fuels and spend money on adaptation. The question is whether philanthropic funds could be used to decommission existing fossil fuel projects or to compensate fossil fuel asset holders for not using these assets, and the conditions under which this would work.

Adapt OECD policy to prohibit ECA finance for fossil fuel projects. Although there is a clear shift in Western policy on export credit for fossil fuels, there are many exceptions to these policies. A stricter policy in this area is desirable and could be scaled up.

9.5. Conclusion and future research

This book set out to tackle the hugely complex question of climate change and the nuances and intricacies of equitably phasing out fossil fuels in a bid to drive effective climate mitigation. This problem is a wicked and tremendously challenging one in nature, both directly and indirectly relying on and concerning not only fossil fuel exploration and production firms, but also their financiers and investors, governments and policymakers, direct and indirect fossil fuel dependents, and civil society in both the Global North and Global South. Nevertheless, we believe that this book draws some new insights that could be used to catalyse vital and niche studies to better grapple with the intricacies of phasing out fossil fuels.

First, the Global North has been so slow to phase out its fossil fuel, and indeed following the war on Ukraine is even investing in new fossil fuel sources, that it has left very little space for the Global South to use its own fossil fuels. Moreover, the Global North continues to invest in fossil fuel in the Global South through a range of different investment options which serves to lock-in the Global South into a fossil fuel economy, while profits are made by investors.

Second, the fundamental contradictions between the short-term gains from using fossil fuels which often outweighs the short and long-term losses from fossil fuel uses tends to motivate political actors and investors with short-term horizons to focus on the short-term gains at the cost of all else. This is often done by using arguments such as supporting the right to development, meeting the needs of the poorest, fiduciary responsibilities, the need to pay back loans, the need to have a higher GDP, etc. All these arguments need to be tackled if the short-term prioritisation is to be reversed. Our analysis showed that export credit agencies (ECAs) from Europe and North America (i.e. the Global North) have aggressively hoisted coal, oil and gas projects across South Africa and Ecuador over the last decades, and plenty of research indicates that other public finance institutions (PFIs) have continued to do so across the entirety of the "Global South" (OCI, n.d.). Now, in the wake of the looming climate emergency, should these PFIs—and the private financiers (i.e. commercial banks) that have jointly hoisted these projects—allocate concessional finance (or perhaps even grants and outright debt relief) to decommission this dirty infrastructure? The Asian Development Bank's Energy Transition Mechanism (ETM) is beginning to dabble with the idea of mobilising a fund that first and foremost is set out to decommission the bulk of Indonesia's coal-fired power infrastructure (Asian Development Bank, 2022) rather than blindly investing in solar PV or wind power development, the latter of which has been the trend since the inception of the Paris Agreement (and arguably even predating it).

Third, social movements are key actors in reversing this. But while social movements generally focus on local challenges and global issues piggyback on these, they are not successful in countries where the civic space is limited, where courts are conservative and where they cannot also address the challenges of other losers—such as labour in the fossil fuel sector and users of fossil fuels. The rise in popularity of the Fossil Fuel Non-Proliferation Treaty Movement, not covered in this book, shows that there is a growing global movement that complements the local social movements.

All in all, the arguments, actors and approaches relevant to a prospective fossil fuel phase-out are ample. This book has drawn from useful findings, but what is most evident is the need for a more innovative and critical paradigm shift at the structural level. NGOs and social movements are playing their part, adopting multidimensional strategies to challenge and oppose fossil projects in their own domains (see Chapters 3 and 4), but governments, policymakers, investors and financiers seem stuck in their "old ways of getting this done." Times have changed, the climate (and, de facto, humankind) is in a state of emergency, and it's time that the world's leaders and the heads of financial markets evolve appropriately, particularly if we are to have any chance whatsoever of "[m]aking finance flows consistent with a pathway towards low greenhouse gas emissions and climate-resilient development" (Paris Agreement, 2015, Article 2.1c).

References

Asian Development Bank. (2022, November 14). Energy Transition Mechanism (ETM). https://www.adb.org/what-we-do/energy-transition-mechanism-etm

Butler, N. (2022, July 14). The impact of the Ukraine war on global energy markets. Centre for European Reform. https://www.cer.eu/insights/impact-ukraine-war-global-energy-markets

Campbell, M. (2022a, March 16). From Angola to the US: Which countries can wean Europe off Russian gas? *Euronews*. https://www.euronews.com/green/2022/03/16/from-angola-to-the-us-which-countries-can-wean-europe-off-russian-gas

Campbell, M. (2022b, April 27). What are Europe's energy alternatives now that Russian gas is off the cards? *Euronews*. https://www.euronews.com/green/2022/04/27/europe-scrambles-to-keep-the-lights-on-as-it-sidelines-russian-gas

Chepeliev, M., Maliszewska, M., & Seara e Pereira, M. F. (2022, May 6). Agricultural and energy importers in the developing world are hit hardest by the Ukraine war's economic fallout. Centre for Economic Policy Research (CEPR). https://cepr.org/voxeu/columns/agricultural-and-energy-importers-developing-world-are-hit-hardest-ukraine-wars

Clifford, C. (2021, November 4). Bezos Earth Fund CEO: Here's how we're giving away $10 billion of Jeff's money. *CNBC*. https://www.cnbc.com/2021/11/04/bezos-earth-fund-ceo-how-were-giving-away-10-billion.html

Coronil, F. (2008). It's the oil, stupid! *Revista: Harvard Review of Latin America, 8*(1), 19–20. https://revista.drclas.harvard.edu/its-the-oil-stupid-an-overview/

Fisher, J. (2022, June 8). Climate change: Ukraine war prompts fossil fuel "gold rush." *BBC News*. https://www.bbc.com/news/science-environment-61723252

Ghiles, F. (2022, July). As North African energy links are redrawn, Italy becomes Europe's southern gas hub. Barcelona Centre for International Affairs (CIDOB). https://www.cidob.org/en/publications/publication_series/notes_internacionals/276/as_north_african_energy_links_are_redrawn_italy_becomes_europe_s_southern_gas_hub

Giridharadas, A. (2018). *Winners take all: The elite charade of changing the world*. Knopf.

Gupta, J., & Lebel, L. (2020a). Access and allocation in earth system governance: Lessons learnt in the context of the Sustainable Development Goals. *International Environmental Agreements: Politics, Law and Economics, 20*, 393–410. https://doi.org/10.1007/s10784-020-09486-4

Gupta, J., & Lebel, L. (2020b). Editorial Access and Allocation in Earth System Governance. *International Environmental Agreements: Politics, Law and Economics, 20*, 197–201. https://doi.org/10.1007/s10784-020-09485-5

Kumar, N., & Levitan, D. (2022, November 3). War in Ukraine is driving a return to coal and a rush to renewables. *Grid News*. https://www.grid.news/story/global/2022/11/01/russias-war-in-ukraine-is-driving-a-rush-for-renewables-and-for-coal-heres-what-that-means-for-the-planet/

Kuzemko, C., Blondeel, M., Dupont, C., & Brisbois, M. C. (2022). Russia's war on Ukraine, European energy policy responses & implications for sustainable transformations. *Energy Research & Social Science, 93*, 102842. https://doi.org/10.1016/j.erss.2022.102842

Muttitt, G., & Kartha, S. (2020). Equity, climate justice and fossil fuel extraction: Principles for a managed phase out. *Climate Policy, 20*(8), 1024–1042. https://doi.org/10.1080/14693062.2020.1763900

Nadig, S. (2022, October 3). Before and after: How the Ukraine crisis has affected Russian oil analysis. *Offshore Technology*. https://www.offshore-technology.com/analysis/before-and-after-how-the-ukraine-crisis-has-affected-russian-oil/

Obisie-Orlu, V. (2022, June 2). Exploring key implications of the Russia–Ukraine war on Africa's energy policy. Africa Policy Research Institute. https://afripoli.org/exploring-key-implications-of-the-russia-ukraine-war-on-africas-energy-policy

OCI [Oil Change International]. (N.d.). Shift the Subsidies Database: Public finance still funding fossils. https://priceofoil.org/shift-the-subsidies/

Paris Agreement. (2015). (Adopted on 12 December 2015, entered into force 4 November 2016), United Nations Treaty Series vol. 3156 C.N.63.2016.

Rammelt, C. F., & Gupta, J. (2021). Inclusive is not an adjective, it transforms development: A post-growth interpretation of inclusive development. *Environmental Science & Policy, 124*, 144–155. https://doi.org/10.1016/j.envsci.2021.06.012

Rempel, A., & Gupta, J. (2021). Fossil fuels, stranded assets and COVID-19: Imagining an inclusive & transformative recovery. *World Development, 146*, 105608. https://doi.org/10.1016/j.worlddev.2021.105608

Ruta, M. (N.d.). How the war in Ukraine is reshaping world trade and investment. World Bank Blogs. https://blogs.worldbank.org/trade/how-war-ukraine-reshaping-world-trade-and-investment

SEI, IISD, ODI, Climate Analytics, CICERO, and UNEP. (2019). *The production gap: The discrepancy between countries' planned fossil fuel production and global production levels consistent with limiting warming to 1.5 °C or 2 °C.* http://productiongap.org/

Tollefson, J. (2022, April 5). What the war in Ukraine means for energy, climate and food. *Nature, 604*, 232–233. https://www.nature.com/articles/d41586-022-00969-9

UNFCCC [United Nations Framework Convention on Climate Change]. (1992). (Adopted in 1992, entered into force 21 March 1994), 1771 UNTS 197 (UNFCCC).

UNFCCC [United Nations Framework Convention on Climate Change]. (2015). Adoption of the Paris Agreement. Report No. FCCC/CP/2015/L.9/Rev.1. Conference of the Parties 21st Session. Paris. https://unfccc.int/sites/default/files/english_paris_agreement.pdf

UNGA [United Nations General Assembly]. (2015, September 25). *Transforming our world: The 2030 Agenda for Sustainable Development.* A/RES/70/1.

UN Global Crisis Response Group on Food, Energy and Finance. (2022, August 3). Global impact of war in Ukraine: Energy crisis. *UN News*. https://news.un.org/pages/wp-content/uploads/2022/08/GCRG_3rd-Brief_Aug3_2022_FINAL.pdf

Vyshnytska, A. (2022, June 7). Russian fossil fuels embargo: "Dictatorships fall when there is enough pressure." Heinrich-Böll-Stiftung. https://eu.boell.org/en/2022/06/07/russian-fossil-fuels-embargo

Index

For Product Safety Concerns and Information please contact our EU representative GPSR@taylorandfrancis.com Taylor & Francis Verlag GmbH, Kaufingerstraße 24, 80331 München, Germany